FOUNDATIONS OF MODERN BIOCHEMISTRY SERIES

Lowell Hager and Finn Wold, editors

ORGANIC CHEMISTRY OF BIOLOGICAL COMPOUNDS*
Robert Barker

INTERMEDIARY METABOLISM AND ITS REGULATION
Joseph Larner

PHYSICAL BIOCHEMISTRY
Kensal Edward Van Holde

MACROMOLECULES: STRUCTURE AND FUNCTION
Finn Wold

SPECIAL TOPICS

BIOCHEMICAL ENDOCRINOLOGY OF THE VERTEBRATES
Earl Frieden and Harry Lipner

THE BIOCHEMISTRY OF GREEN PLANTS
David W. Krogmann

AN INTRODUCTION TO
BIOCHEMICAL REACTION MECHANISMS
James N. Lowe and Lloyd L. Ingraham

EXPERIMENTAL TECHNIQUES IN BIOCHEMISTRY
J. M. Brewer, A. J. Pesce, and R. B. Ashworth

*Published jointly in Prentice-Hall's *Foundations of Modern Organic Chemistry Series.*

AN INTRODUCTION TO BIOCHEMICAL REACTION MECHANISMS

AN INTRODUCTION TO BIOCHEMICAL REACTION MECHANISMS

JAMES N. LOWE

Associate Professor of Chemistry
The University of the South
Sewanee, Tennessee

LLOYD L. INGRAHAM

Professor of Biophysics
University of California, Davis
Davis, California

Prentice-Hall, Inc., Englewood Cliffs, New Jersey

Library of Congress Cataloging in Publication Data

LOWE, JAMES N

An introduction to biochemical reaction mechanisms.

(Foundations of modern biochemistry series)

Includes bibliographies.

1. Biological chemistry. 2. Chemistry, Organic.
I. Ingraham, Lloyd L., joint author. II. Title.

QH345.L78 1974 574.1'924 73-22471

ISBN 0-13-478545-2

Printed in the United States of America

10 9 8 7 6 5 4 3 2 1

PRENTICE-HALL INTERNATIONAL, INC., *London*
PRENTICE-HALL OF AUSTRALIA, PTY. LTD., *Sydney*
PRENTICE-HALL OF CANADA, LTD., *Toronto*
PRENTICE-HALL OF INDIA PRIVATE LIMITED, *New Delhi*
PRENTICE-HALL OF JAPAN, INC., *Tokyo*

To our wives, MARTHA and IDA

CONTENTS

PREFACE

Biochemistry has undergone some drastic changes from the original studies of "ferments" to modern day. Biochemistry has expanded in directions toward both biology and chemistry. The more chemical aspects of biochemistry, including mechanism of reactions, was at first reserved for the domain of the graduate student in biochemistry. Today, with more chemically trained biochemists and chemists showing a great interest in biochemical problems, the chemical aspects of biochemistry are of interest at the undergraduate level.

The purpose of this book is to fulfill the need to add more basic chemistry to the undergraduate curriculum in biochemistry. This book can be used as a supplement in a biochemistry course or an organic course as well as for independent study. Many instructors may wish to add some biochemical examples to an organic course, for today most of the students in organic chemistry courses are not chemists but potentially biologists or physicians who are interested in how organic chemistry applies to their interests.

The authors have been impressed with the chemical perfection of biological systems. Biochemistry is chemistry at its ultimate degree of sophistication.

Coenzymes are well-designed molecules that perform certain chemical tasks. As chemists, we are awe struck with the beauty and perfection of these molecules. Science can be beautiful, and biochemistry is the ultimate in chemical beauty.

We are grateful to the series editors and to the students at Sewanee and at Davis for their interest and enthusiasm as we developed and prepared this material. We thank Barbara Hart for her skillful work in the typing of this manuscript.

JAMES N. LOWE
LLOYD L. INGRAHAM

1 ENZYMES

1.1 INTRODUCTION

A living system is a remarkable chemical system. Many step syntheses of large molecules from small molecules proceed in very high yields. Chemical and light energy is converted into energy for synthesis, for locomotion, for transport of molecules and ions against concentration gradients, for creating and maintaining electrical fields, and into heat. In a simple organism, thousands of different reactions occur at the same temperature without strong acids and bases and without the special solvents found in chemical laboratories.

These reactions are made possible by chemical catalysts called *enzymes.* A given enzyme catalyzes one or more reactions very effectively. Reactants in enzyme catalyzed reactions are called *substrates.* An enzyme may be specific for a given substrate—such as the hydrolysis of pyrophosphate ion to give two orthophosphate ions—or it may be general for a class of substrates—such as the addition of properly activated amino acids to a growing peptide chain.

Enzymatic reactions are regulated. An enzyme may bond to a small nonsubstrate molecule (including products), and this now modified enzyme may be turned on or turned off completely, or it may catalyze the reaction of substrate at a different velocity.

To understand the chemistry of biological systems, we must understand the nature of enzyme catalysis. As chemists, we would like to identify the relation of enzyme structure to the function of catalysis.

One approach is to identify factors common to several enzymes. In part, this task is one of finding the connection between mechanisms of enzymatic reactions and the mechanisms of organic reactions. Solvent effects and acid and base catalysis are expected to play corresponding roles. An understanding of features of enzymatic catalysis is validated by the design and testing of model chemical systems. As a by-product of such studies our understanding of enzymatic mechanisms may enable us to design better organic catalysts.

A second approach to understanding enzymatic catalysis is to attempt to understand intimately the function of a particular enzyme. Just as time lapse photography or videotape replay may let us observe the form displayed by an athlete, pictures of enzymes and of enzymic substrate complexes are accumulating from the work of X-ray crystallographers. Unfortunately for our study, but fortunately for living organisms, enzyme complexes with reactive substrates react. Pictures can only be taken with poor or incomplete substrates. While many frames may be missing from our picture, an appreciation of the cooperative interaction of different groups on the enzyme with the substrate is emerging. This view of enzyme mechanisms will not be discussed in this book; it is readily available elsewhere.

There are other ways to attempt to understand enzymatic catalysis. Models for mechanisms can be tested against kinetic predictions. Reaction velocities can be measured as a function of enzymic concentration, of substrate or inhibitor concentration, of pH, and of temperature. Binding constants for the binding of the substrate to the enzyme can be measured. Isotopes may be used to determine whether a given bond is cleaved in the rate determining step and to trace the path of any given atom. The strategies are in general similar to, but not identical in detail with, those used by organic chemists.

In the remainder of this chapter, the structure of enzymes will be presented to identify the environment of biochemical transformations. Certain features of enzyme structure will be abstracted and identified, and their role in catalysis will be assessed by consideration of evidence obtained from model systems.

1.2 THE STRUCTURE OF ENZYMES

Enzymes constitute the class of proteins that catalyze reactions. Other proteins play structural or transport roles in cells. Proteins are formally derived by the condensation polymerization of α-amino acids (1.1). The covalent sequence of the linear chain is referred to as the primary structure of proteins. Proteins are linear–amino groups and carboxyl groups of side chains are not bonded to other amino acid residues. Proteins range in molecular weight from about 6,000 to over 1,000,000. Very high molecular weight proteins consist of associated shorter chains (quaternary structure). Proteins differ in their amino acid composition and sequence, degree of polymerization, shape, physical, and chemical properties.

$$n\ \overset{+}{N}H_3CHRCO_2^- \rightarrow \overset{+}{N}H_3CHR\overset{O}{\overset{\|}{C}}{-}(NHCHR\overset{O}{\overset{\|}{C}})_{n-2}{-}NHCHRCO_2^- + (n-1)\,H_2O \tag{1.1}$$

As in other biopolymers, the number of different monomers found in proteins is limited. The structures of the twenty-one amino acids regularly found in proteins are given in Table 1.1. The side chains of these amino acids range from polar to non-polar, from acidic to basic, and present a variety of organic functional groups. In Table 1.1, the amino acids are classified as acidic, basic, and neutral based upon the side chain functional group.

TABLE 1.1

	Amino Acid	*Structure at pH7*
Uncharged nonpolar side chain	Glycine	$H{-}CH(\overset{+}{N}H_3)CO_2^-$
	Alanine	$CH_3{-}CH(\overset{+}{N}H_3)CO_2^-$
	Valine	$(CH_3)_2CH{-}CH(\overset{+}{N}H_3)CO_2^-$
	Leucine	$(CH_3)_2CHCH_2{-}CH(\overset{+}{N}H_3)CO_2^-$

TABLE 1.1 (contd.)

	Amino Acid	*Structure at pH7*
	Proline	H_2C—CH_2—$CHCO_2^-$ (ring: H_2C—H_2C—$\overset{+}{N}H_2$—$CHCO_2^-$)
	Phenylalanine	C_6H_5—CH_2—$CH(\overset{+}{N}H_3)CO_2^-$
	Tryptophan	indolyl(N—H)—CH_2—$CH(\overset{+}{N}H_3)CO_2^-$
Uncharged nonpolar side chain	Methionine	$CH_3SCH_2CH_2$—$CH(\overset{+}{N}H_3)CO_2^-$
Uncharged polar side chain	Serine	$HOCH_2$—$CH(\overset{+}{N}H_3)CO_2^-$
	Threonine	$CH_3CH(OH)$—$CH(\overset{+}{N}H_3)CO_2^-$
	Cysteine	$HSCH_2$—$CH(\overset{+}{N}H_3)CO_2^-$

TABLE 1.1 (contd.)

	Amino Acid	*Structure at pH7*
	Tyrosine	$HO-C_6H_4-CH_2-CH(\overset{+}{N}H_3)CO_2^-$
	Asparagine	$H_2N-C(=O)CH_2-CH(\overset{+}{N}H_3)CO_2^-$
	Glutamine	$H_2N-C(=O)CH_2CH_2-CH(\overset{+}{N}H_3)CO_2^-$
Acidic side chain (negatively charged at pH7)	Aspartic Acid	$^-O-C(=O)CH_2-CH(\overset{+}{N}H_3)CO_2^-$
	Glutamic Acid	$^-O-C(=O)CH_2CH_2-CH(\overset{+}{N}H_3)CO_2^-$
Basic side chain (positively charged near pH7)	Lysine	$\overset{+}{N}H_3CH_2CH_2CH_2CH_2-CH(\overset{+}{N}H_3)CO_2^-$

TABLE 1.1 (contd.)

Amino Acid	*Structure at pH7*
Arginine	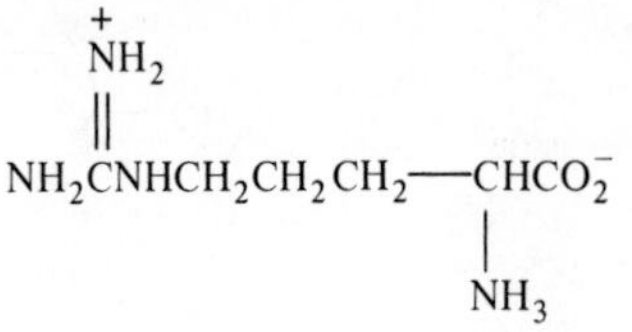 $NH_2\overset{\overset{+}{NH_2}}{\overset{\Vert}{C}}NHCH_2CH_2CH_2—\underset{\underset{NH_3}{\vert}}{C}HCO_2^-$
Histidine (at pH 6.0)	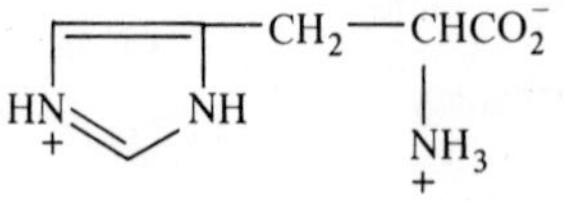 $\overset{+}{HN}$ imidazole ring NH $—CH_2—\underset{\underset{\overset{+}{NH_3}}{\vert}}{C}HCO_2^-$

Glycine, the simplest amino acid is achiral. All other amino acids have the same configuration at the chiral center; this configuration (1.2) is designated as *S* according to the Cahn-Ingold-Prelog convention or L-configuration relative to D-glyceraldehyde.

(1.2)

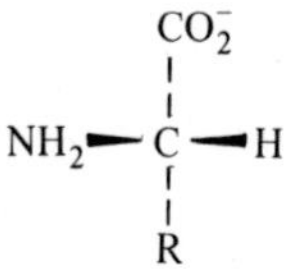

$$NH_2—\overset{\overset{CO_2^-}{\vert}}{\underset{\underset{R}{\vert}}{C}}—H$$

The amide bonds between amino acid residues are referred to as the peptide linkages, and proteins are called *polypeptides*. Resonance of the peptide (1.3) is

(1.3)

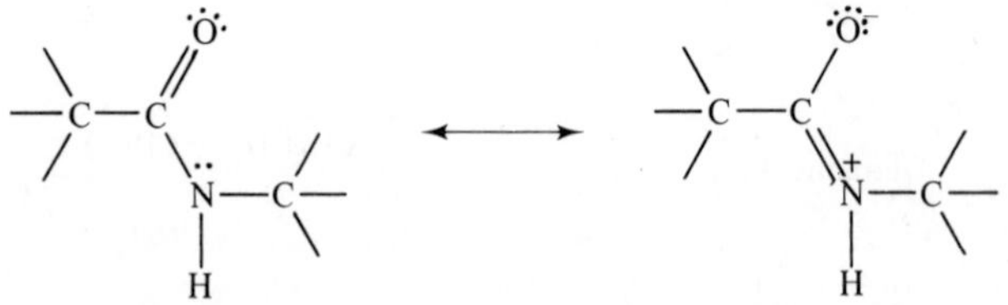

$$>C—C(=\ddot{O})—\ddot{N}(H)—C< \longleftrightarrow >C—C(—\ddot{O}^-)=\overset{+}{N}(H)—C<$$

important for the conformation of proteins. Because of the double bond character of the carbon-nitrogen bond, the six atoms shown in (1.3) lie in a plane. To minimize nonbonded interactions, adjacent peptide linkages lie in different planes. The particular dihedral bond angles that occur between adjacent amino acids depend on the particular side chains. To a lesser extent, interactions with more distant groups influence geometry. The influence of nonbonded repulsions to the configurational energy is as important as the better publicized hydrogen bonding discussed below.

Hydrogen bonding is important in the structure of polypeptides as it is in simple amides. (Acetamide is a solid but ethyl acetate is not.) Structures that maximize hydrogen bonding are generally found in proteins. The peptide NH may be hydrogen bonded to a side chain base, to water, or to the carbonyl oxygen of another peptide bond. Similarly, a side chain hydroxyl or ammonium group, water, or a peptide NH may hydrogen bond to the carbonyl oxygen.

Certain twists on foldings of the peptide chain maximize hydrogen bonding of peptide groups in the same chain. Two such structures are commonly found in enzymes, the α-helix and the β-pleated sheet. These structures feature near minimum energy dihedral angles between adjacent planar peptide linkages and maximum hydrogen bonding between peptide groups. The additive energies of several hydrogen bonds (about 7 kcal/mole/hydrogen bond) provide the energy to maintain the three-dimensional structure. The α-helix and β-pleated sheet are referred to as the secondary structure of proteins.

In the α-helix (1.4) hydrogen bonds form between groups that are 3.6 residues apart. In contrast, the β-pleated sheet (1.5) features hydrogen bonds between antiparallel segments of a chain or between different chains. For a single chain to exist in a β-pleated sheet, there must be turns in which internal hydrogen bonding is not maximized.

(1.4)

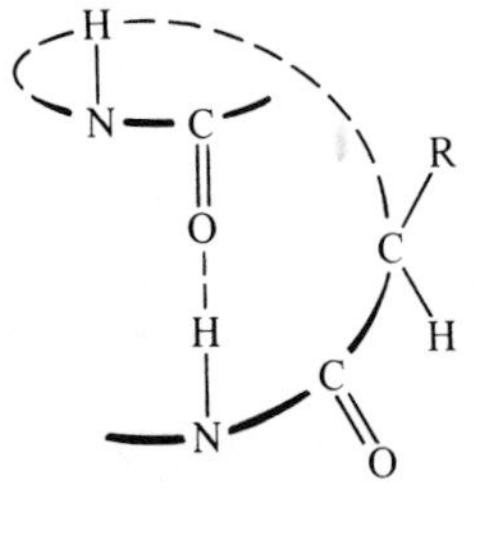

(1.5)

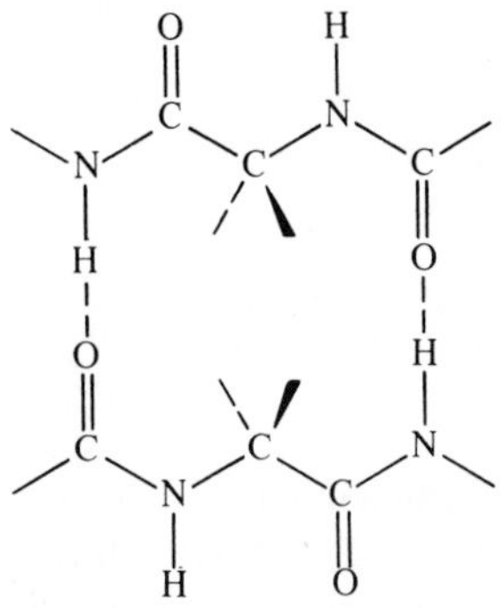

The three-dimensional structure of protein molecules is referred to as the tertiary structure of proteins. X-ray structures reveal that most enzymes are compact, containing little solvent. The structures show regions that are helical, regions that correspond to sheets, and other less regular regions. Groups formally far apart on the backbone sequence may be close together in space. Some of the three-dimensional structure results from disulfide linkages between cysteine side chains. These form after the linear polypeptide has been synthesized. The protein must have folded in such a way that the sulfhydryl groups being oxidized have been brought close to one another.

Aliphatic and aromatic side chains tend to be found in the interior regions of an enzyme; polar groups and particularly ionic groups are most frequently on the surface. The surface of an enzyme is largely hydrophilic; the interior is hydrophobic. The relative solubility of side chains in one another and in water is another major contributing factor to enzymatic shape (and to the binding of small molecules). The three-dimensional folding of a peptide chain is referred to as the tertiary structure of proteins.

1.3 THE ACTIVE SITE OF ENZYMES

The isolation of a single variable is often difficult when working with an enzyme. Experiments may be difficult to interpret or are capable of more than one interpretation.

For example, a particular enzyme may be deactivated by the chemical modification of a cysteine side chain by reaction with iodoacetamide (1.6). Does the loss of activity mean that the alkylated sulfhydryl group directly participated in the reaction catalyzed? Or has the enzyme assumed a different, unreactive conformation even though the sulfhydryl group played no role in the enzymatic activity? On the basis of a single experiment, these effects cannot be distinguished. Evidence for enzyme mechanisms is frequently circumstantial.

$$\text{enzyme—CH}_2\text{SH} + \text{ICH}_2\text{C(=O)NH}_2 \rightarrow \text{enzyme—CH}_2\text{SCH}_2\text{C(=O)NH}_2 + \text{H}^+ + \text{I}^- \quad (1.6)$$

Complementary evidence using a variety of probes is necessary in establishing the mechanism of enzymes.

The dilemma of whether a given modification involves a chemically active site or whether it involves a conformational change is difficult. Experiments could have been interpreted by two models. In one, the enzyme is a template with a preformed site suitable for binding the substrate and catalyzing its reaction (1.7): a modification of enzyme structure that greatly alters the enzymatic activity affected the active site, whereas a modification of the enzyme that has little effect on reactivity took place elsewhere. The participation of particular functional groups at a catalytic site on the enzyme could be tested by selective modification of the enzyme and assays of activity. The alternate hypothesis was not operational in determining mechanisms. The enzyme-substrate relation was

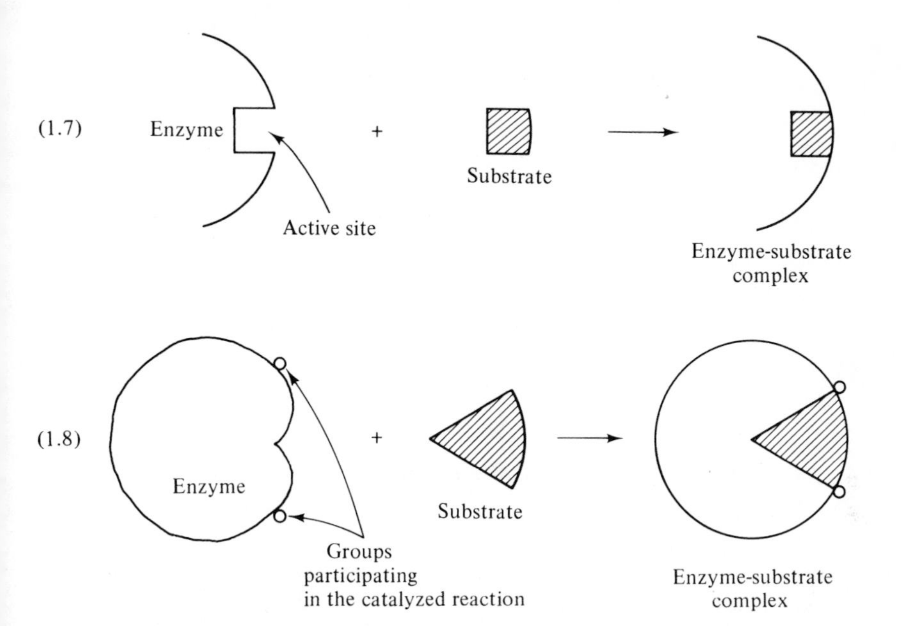

viewed as that of "lock and key" (1.8). When the enzyme formed a complex with the substrate, groups came together to catalyze the reaction. Chemical modifications changed the possible geometries in unpredictable ways according to this model. Attributing changes of structure and reactivity to an active site was not justified by this hypothesis.

By the use of X-ray crystallography, "eye witness" support for the active site hypothesis has been obtained. Again the evidence is partly circumstantial. A substrate at the active site reacts; transition states do not form crystals. However, poor substrates or incomplete substrates may bind and may be studied in favorable circumstances.

Lysozyme, an enzyme isolated from tears and from egg white, dissolves the cell walls of certain bacteria. Its substrate is a polysaccharide consisting of alternating units, ABAB. . . . The crystalline enzyme resembles an open clam with a top, a bottom, and a crease. A crystalline complex of the enzyme with a tetrasaccharide ABAB has also been studied. This partial substrate is bonded in the crease. Models indicate that two more segments would fit in the crease, but the reactive site for hydrolysis would cleave the bond between the fourth and fifth residues (1.9). It is significant that the enzyme and the complex of the enzyme with the incomplete substrate are similar (the changes in geometry are small). There is an active site (an enzyme is not amorphous), although changes in conformation may occur and may be important in the mechanism. Work with other enzymes supports the generality of this model; in no case is it contradicted.

(1.9)

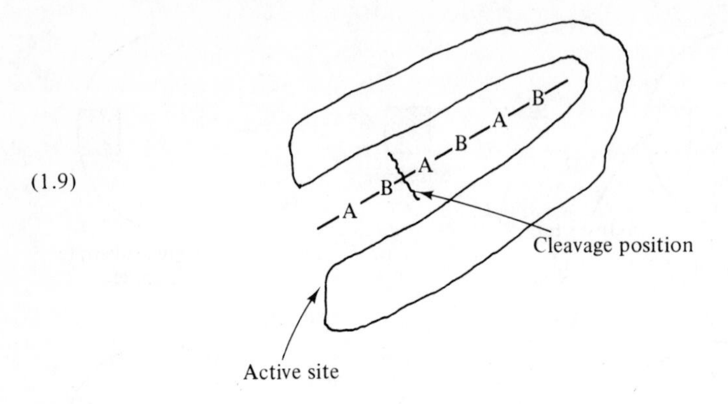

1.4 FACTORS INVOLVED IN CATALYSIS BY ENZYMES

In this section, we will take the view that enzymes are large organic (or organometallic) catalysts. Their efficiency is one of design optimization; their construction and function do not involve new principles. In this survey, the discussion will be brief–particular mechanisms will be discussed in later chapters. The viewpoint is that of the organic chemist. Analogies to simpler model systems will be used to provide a way to factor variables for experimental study.

A second-order homogeneous reaction, such as the S_N2 reaction given in (1.10),

$$I^- + CH_3CHCl(CH_2)_5CH_3 \rightarrow CH_3CHI(CH_2)_5CH_3 + Cl^- \quad (1.10)$$

occurs by a biomolecular mechanism in homogeneous solution. The reaction rate is related to an activation energy. This activation energy includes an enthalpy that depends on the strength of the bond to be broken, upon the nucleophilicity of the attacking group, and upon the ability of the solvent to solvate the transition state. The activation energy also includes an entropic term; the reaction takes place by a Walden inversion at the chiral center. Only collisions of the proper geometry can lead to products.

Were this reaction to occur on an enzyme, we could identify passive ways in which the reaction can be speeded. A plausible mechanism is (1.11).

(1.11)

(a) $A + \text{Enzyme} \rightarrow A \cdot \text{Enzyme}$

(b) $A \cdot \text{Enzyme} + B \rightarrow [A\ \text{Enzyme} \cdot B]$

(c) $[A \cdot \text{Enzyme} \cdot B] \rightarrow C + D + \text{Enzyme}$

Steps (b) and (c) may not be distinct. Because termolecular collisions are exceedingly rare, at least one of the substrates must bind to the enzyme for reaction to occur. Whether or not the second substrate binds, the collisions are no longer random. The entropy of reaction has been changed. The enzyme has local solvation properties that govern the solvation energy for the reaction.

In the organic laboratory, we may wish to convert a carboxylic acid to an ester. Common strategies include the acid catalyzed reaction of the acid with an alcohol and the preparation of a reactive acid derivative such as an acid chloride, followed by reaction with an alcohol. An enzyme might catalyze such a reaction in similar ways. It may provide acid and base catalysis for reactions, or it may react with a substrate to form an intermediate that can react to form product.

Factors of ordered concentration (approximation) of reactants, solvation, acid and base catalysis, and covalent participation will be treated individually. A fifth factor, one of design optimization, will also be treated. That is, an enzyme may be designed to stabilize the transition state for a reaction more than it does the reactants.

Approximation

Small molecules may bond strongly to enzymes. Dissociation constants for enzyme-substrate complexes and enzyme-inhibitor complexes vary over several orders or magnitude. Values as low as 1×10^{-15} have been measured. By selectively binding reactants, enzymes furnish a local concentration of reactants that approaches the concentration of solvent (55 *M* for water).

A more subtle effect is the positioning of reactants at the active site. Organic chemists have found that the relative positions of functional groups in a molecule can determine the product obtained in a reaction (1.12). This product distribution results from the competition between inter- and intra-molecular reactions. The intramolecular reaction is fast when the reacting groups are close, as is the case for five- and six-member ring formation. For larger ring formation, the reacting groups are not often close in low energy conformations; hence, the intermolecular reaction predominates.

(1.12)

$$HOCH_2(CH_2)_n\,COOH \xrightarrow{\text{heat}} \begin{cases} (CH_2)_{n+1}\text{—C(=O)—O (cyclic lactone)} & n = 2,3 \\ -(O(CH_2)_{n+1}\overset{O}{\overset{\|}{C}})-_x & n > 3 \end{cases}$$

TABLE 1.2 RELATIVE RATE OF LACTONE FORMATION

Compound	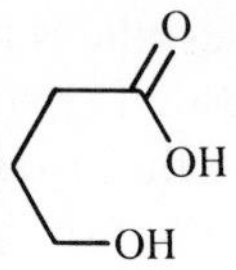 (4-hydroxybutanoic acid)	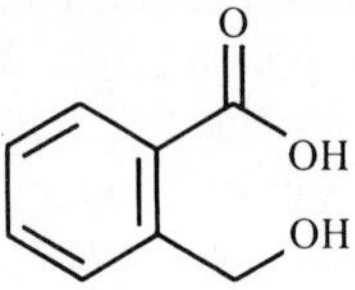 (2-(hydroxymethyl)benzoic acid)	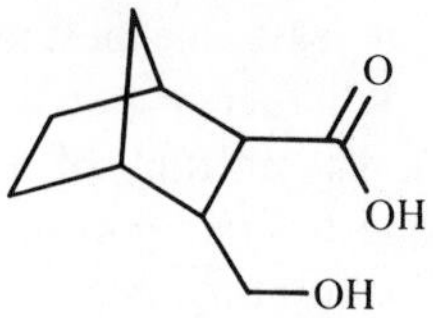 (bicyclic hydroxy acid)	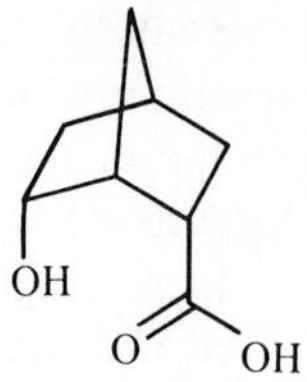 (bicyclic hydroxy acid)
Relative rate	1	3.9	84	13×10^3

Modified from D. R. Storm and D. E. Koshland, Jr., *Proc. Nat. Acad. Sci. U.S.*, **66,** 445 (1970)

When the geometry of the molecule is fixed so that groups are held close together, the intramolecular reactions are more rapid. The relative rates of lactone formation for various hydroxyacids are given in Table 1.2. As this comparison indicates, the positioning of groups that is possible on an enzyme surface can powerfully accelerate reactions.

Solvent Effects

Solvent effects occur over short distances. The wide variety of side chains of amino acids makes it possible for an enzyme to have a wide range of environments at the active site. We will look at one measure of solvent effects, namely, spectral shifts.

Nicotinamide adenine dinuoleotide (NAD^+) is part of a common redox pair. The reaction of NAD^+ to NADH is given in (1.13).

(1.13) $$\text{(pyridinium ring with } CONH_2\text{, } N^+\text{–R)} + 2e^- + H^+ \rightarrow \text{(dihydropyridine ring with } H, H\text{, } CONH_2\text{, } \ddot{N}\text{–R)}$$

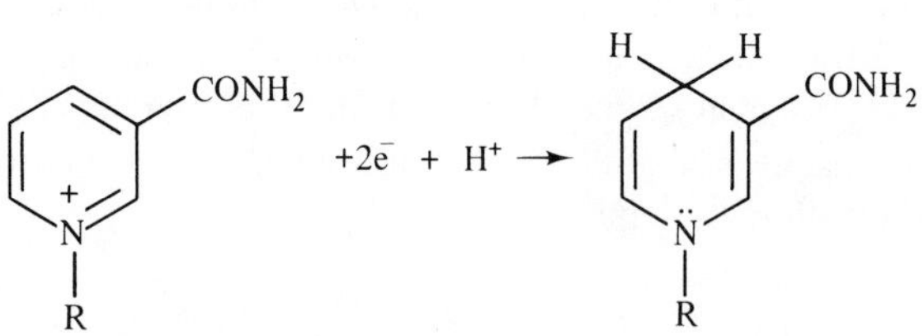

Because NAD^+ is charged and NADH is not, the half-reaction should be shifted to the right when the NAD^+ or NADH is placed in a less polar environment. The UV spectrum of NADH has two bands. One corresponds to a $\pi \rightarrow \pi^*$ excitation. The longer wavelength band (340 nm in water) is very sensitive to solvent effects. This is the excitation of a lone pair (1) electron to a π^* orbital (1.14). Polar solvents stabilize the excited state relative to the ground state while nonpolar solvents raise the energy of the excited state relative to the ground state. This corresponds to a blue shift (shorter wavelength) in nonpolar solvents. For the reduced pyridine ring, when R is isopropyl, λ_{max} is 340 nm in ether; in

(1.14) $\xrightarrow{h\nu}$ (structures: O, NH_2, N, R → O, NH_2, N^+, R)

water it is 357 nm. For NADH in water, λ_{max} is 340 nm; when complexed to the enzyme, alcohol dehydrogenase, the value is 325 nm. Thus the environment on the protein is quite hydrophobic.

An alternate explanation can be given for the spectral shift in enzyme bound NADH. The dipolar photoexcited state could be destabilized in a polar medium if there were a negative charge near the oxygen atom or a positive charge near the nitrogen atom. Such a fixed distribution of charges would be unlikely in solution, but is quite possible on an enzyme surface. It is also quite possible that a combination of selective solvation and selective charge distribution is used by enzymes.

The titration of amino acid side chains of proteins also gives evidence for widely different solvent effects. The pK_a of a lysine side chain may vary by as many as two units, depending on the protein.

Solvent effects give us a mechanism for providing a gradation of reactivities. For instance, the potential of the NAD^+–NADH redox couple can be made to vary by the solvation at its binding site. On different enzymes, this value will be different. This variation increases the utility of the half-reaction in biological systems. In a similar way, the use of solvent effects to increase or decrease the strength of an acid or base may be important for the catalysis of chemical reactions.

Thus far we have looked at evidence for varied solvent environments on enzymes. The potential dramatic influence of solvation on reactions can be seen in reactions of dibasic acids (Table 1.3). The ratio of the first acid dissociation constant to the second is very dependent on the effective dielectric constant of

TABLE 1.3 RATIO OF FIRST AND SECOND IONIZATION CONSTANTS OF DIBASIC ACIDS

Acid	COOH | CH_2 | COOH	COOH | $CH_3CH_2CCH_2CH_3$ | COOH	COOH | CH_2 | CH_2 | COOH	COOH | CH_3CCH_3 | CH_3CCH_3 | COOH
K_1/K_2	734	121,000	19.2	6130

the medium. In the alkyl substituted malonic and succinic acids, nonpolar side groups exclude water, allowing the second carboxyl group to feel the influence of the charge of the first to a greater extent. The energy of the dianion is raised, making it harder for a base to remove the second proton and increasing the ratio K_1/K_2.

Covalent Catalysis

This topic will be treated in considerable detail in later chapters, in examples of reactions. At present we may note that enzymes have carboxyl, hydroxyl, histadyl, and sulfhydryl groups on side chains. These are all nucleophiles. Coenzymes, introduced in Section 1.6, can also covalently participate in reactions. A reason for the formation of such intermediates is that enzymes can use covalently bonded intermediates to maximize acid or base catalysis of reactions. In this section, evidence for such intermediates will be given; in the next section, the catalytic advantage of such intermediates will be discussed.

Chymotrypsin is an extensively studied digestive enzyme. It catalyzes the hydrolysis of a wide variety of substrates – esters as well as amides. Either by using substrates that react slowly, or by studying the reaction at a pH where the reaction is sluggish, a variety of evidence for the formation of an acyl serine intermediate (1.15) has been accumulated. Using p-nitrophenyl acetate, the

$$\mathrm{RC(=O)OR'\ (NHR')} + \underbrace{\mathrm{CH_2OH}}_{\text{enzyme}} \longrightarrow \tag{1.15}$$

$$\mathrm{RC(=O)O{-}}\underbrace{\mathrm{CH_2}}_{\text{enzyme}} + \mathrm{R'OH\ (R'NH_2)}$$

release of p-nitrophenol can be followed spectrometrically. If the formation of the covalent intermediate is fast, and the subsequent hydrolysis of the acetylenzyme is slow (1.16), the initial rate of p-nitrophenol formation should be greater than the steady state speed after a stoichiometric amount of ester has reacted. This is the case. This is because a phenol is a better leaving group than an alcohol. Competing nucleophiles can be used to provide evidence for an intermediate (the tertiary butyl cation can be trapped with a variety of

$$\underbrace{\mathrm{RC}(=\mathrm{O})-\mathrm{O}-\mathrm{CH_2}}_{\text{enzyme}} + \mathrm{H_2O} \longrightarrow \underbrace{\mathrm{CH_2OH}}_{\text{enzyme}} + \mathrm{CH_3C}(=\mathrm{O})\mathrm{OH} \tag{1.16}$$

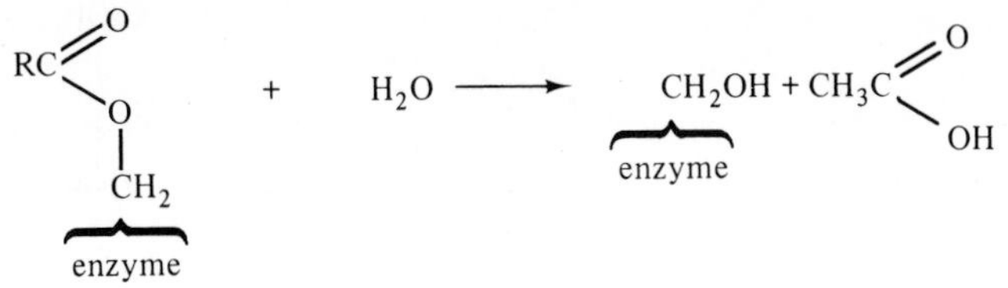

nucleophiles). With amides as the substrate, the second step–the hydrolysis of an ester–is faster than the formation of the ester from the less reactive amide. In solutions of hydroxylamine, the rate of reaction of the amide, *N*-acetyltyrosine-*p*-nitroanalide(1.17a), is constant, but the product varies with hydroxyl amine concentration (1.17b).

(1.17)

(a) $\mathrm{HO{-}C_6H_4{-}CH_2CH(NH{-}C(=O){-}CH_3)C(=O)NH{-}C_6H_4{-}NO_2}$

Bond cleaved in hydrolysis (arrow to the NH–C(=O) bond)

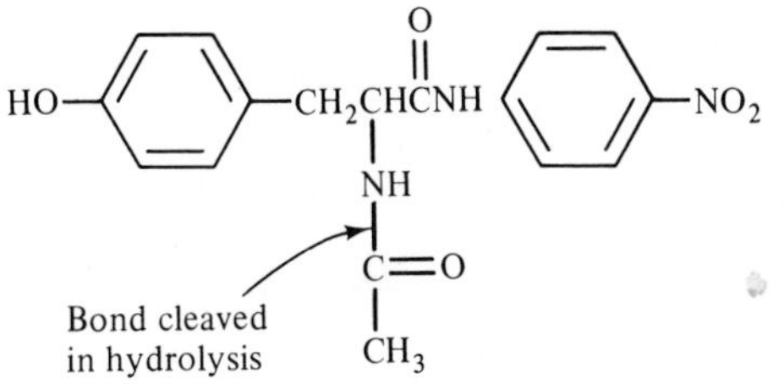

Acetyl—NA + Enzyme → NA + Acetyl—Enzyme

(b) Acetyl—Enzyme $\xrightarrow{H_2O}$ Acetic Acid + Enzyme

Acetyl—Enzyme $\xrightarrow{H_2NOH}$ N-Hydroxyacetamide + Enzyme

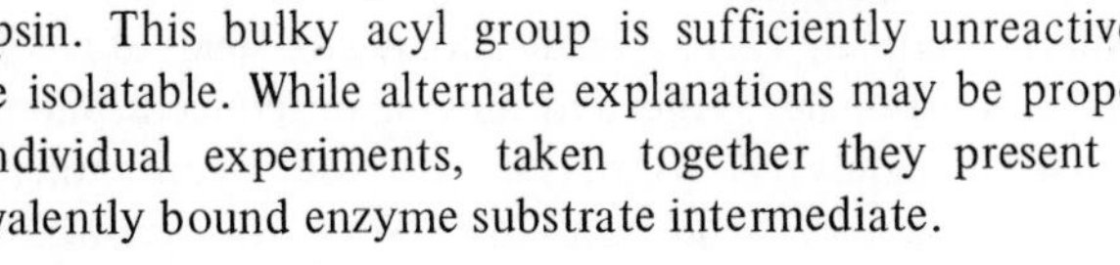

Still more powerful evidence is provided by the crystallization of trimethylacetyl chymotrypsin. This bulky acyl group is sufficiently unreactive in the second step to be isolatable. While alternate explanations may be proposed for each of these individual experiments, taken together they present forceful evidence for a covalently bound enzyme substrate intermediate.

Acid and Base Catalysis

The carbon-oxygen bond of esters is very strong. To cleave esters in the organic laboratory we must use a strong acid (1.18) or a strong base (1.19). The reactive intermediates in these reactions are not available in neutral solution. For an enzymatic reaction, in contrast to the above reactions, the transfer of protons may be as important as making and breaking other bonds.

$$H^+ + R{-}C(=O)OR' \longrightarrow R{-}C(=\overset{+}{O}H)OR' \rightarrow \rightarrow R{-}C(OH)(OH)(\overset{+}{H}OR')$$

$$(1.18) \qquad \longrightarrow R{-}C(=\overset{+}{O}{-}H)OH + HOR'$$

$$\downarrow$$

$$R{-}C(=O)OH + H^+$$

$$HO^- + R{-}C(=O)OR' \longrightarrow R{-}C(O^-)(OR')(OH) \longrightarrow$$

$$(1.19)$$

$$R{-}C(=O)OH + {}^-OR' \longrightarrow R{-}C(=O)O^- + HOR'$$

The organic chemist can take a bottle of strong acid or base off the shelf to hydrolyze an ester, but these are not available to a cell which must remain close to neutrality so as to prevent damage to the cell as a whole. Table 1.4 lists amino acid side chains together with their approximate pk_a's in proteins. These are the acids and bases available for enzymatic reactions.

An intermediate in (1.18) is the conjugate acid of an ester. Its concentration is proportional to the hydronium ion concentration. Reaction (1.18) is specific-acid catalyzed. (Catalysis by the hydronium ion–but not by other acids–is *specific acid catalysis.*) When the hydrolysis of the amide acetylimidazole (1.20) is studied at pH's between 7 and 8, it was found that the rate of reaction depended on the concentration of the buffer. Either the imidazolium ion or imadazole of the buffer catalyzed the reaction. If a reaction is catalyzed by other bases in addition to hydroxide ion, the catalysis is referred to as *general base catalysis.* The corresponding acid catalysis is *general acid catalysis.* The acids and bases on enzymes can participate in general acid and base catalysis of reactions.

TABLE 1.4 ACIDIC AND BASIC SIDE CHAINS OF AMINO ACIDS

Amino Acid	*Acidic Group*	*Approximate pK_a in a protein*
Glutamic Acid Aspartic Acid	—COOH	4.7
Histadine	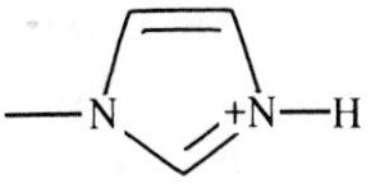 —N(imidazolium ring)+N—H	6.5
Cysteine	—SH	9.0
Tyrosine	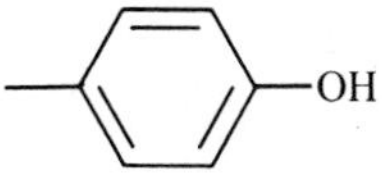 —C_6H_4—OH	10.0
Lysine	$—NH_3^+$	10.2
Arginine	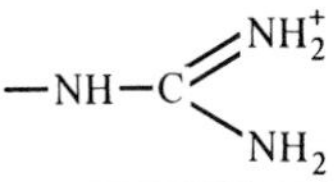 $—NH—C(=NH_2^+)NH_2$	12.0

(1.20)
$$CH_3\overset{O}{\overset{\|}{C}}—N\text{(imidazole)}N + H_2O \longrightarrow CH_3CO_2^- + HN\text{(imidazole)}N + H^+$$

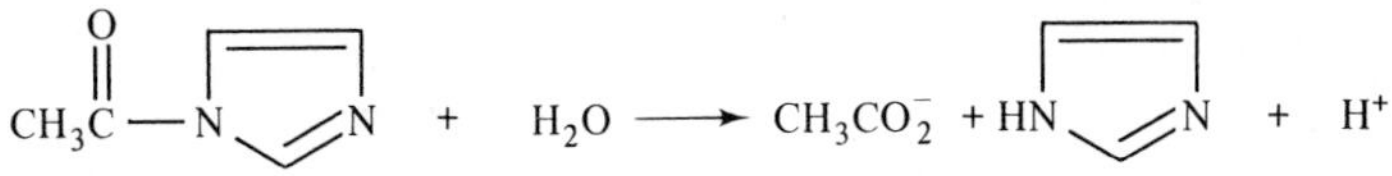

The rates of formation and hydrolysis of the acyl serine bond on chymotrypsin are highly pH dependent near pH 7. As the pH is increased the rate of reaction increases. We can see from Table 1.4 that the concentration of free imidazole increases with this increase in pH. This suggests that an imidazole at the active site provides general base catalysis of the reactions.

Support for the involvement of particular amino acids in a reaction is often obtained by the chemical modification of a particular residue. This was done with an imidazole of chymotrypsin. The chloroketone (1.21) resembles

(1.21)
$$C_6H_5—CH_2\underset{\underset{\underset{SO_2—C_6H_4—CH_3}{|}}{NH}}{\underset{|}{C}}H\overset{O}{\overset{\|}{C}}CH_2Cl$$

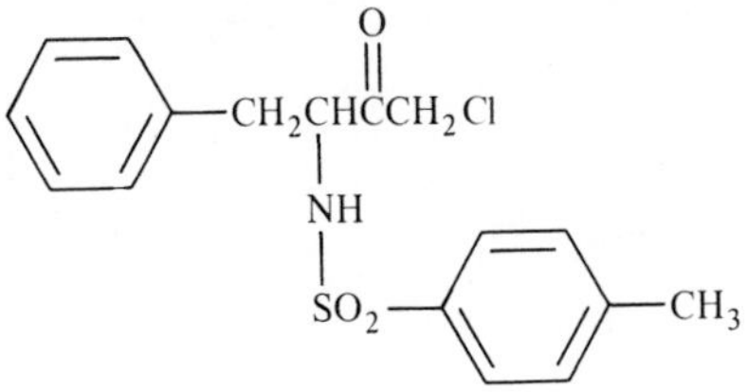

chymotrypsin substrates sufficiently to bind at the active site. For short periods of time, the ketone competes with other substrates competitively and reversibly. In a slow reaction, imidazole is alkylated in a nucleophilic displacement of chloride. This reaction irreversibly deactivates the enzyme.

A scheme that accounts for the base catalysis by imidazole is given in (1.22). The imidazole removes a hydrogen bonded proton from the water as the water attacks. As the base removes the proton from water, the nucleophile becomes more like a hydroxide ion. The rate increases. The resulting imidazolium ion then catalyzes the breaking of the acyl serine bond by hydrogen bonding to the serine oxygen and transferring a proton as the carbon-oxygen bond is cleaved.

(1.22)

Because water is both a good acid and a good base, it is necessary to go to other solvents to study the importance of combined acid and base catalysis. The rate of hydrolysis of 2,3,4,6-tetramethylglucofuranoside in benzene was studied with various catalysts (1.23). This reaction might be expected to be subject to both acid and base catalysis. Both phenol and pyridine catalyze the reaction.

(1.23)

However, when the acidic and basic group were combined in one molecule (2-pyridone), the catalysis was much more effective (1.24). The combined acid and base transfers interconvert the tautomers of this catalyst. (A quantitative comparison is difficult, for the number of molecules involved in the reaction changes when a bifunctional catalyst is used.)

(1.24)

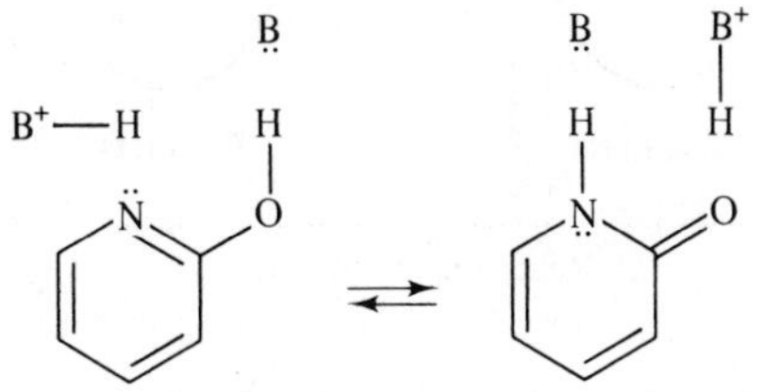

Compared to 2-pyridone, 2-aminophenol is a very poor catalyst. The corresponding proton shifts result in an unfavorable charge separation (1.25).

(1.25)

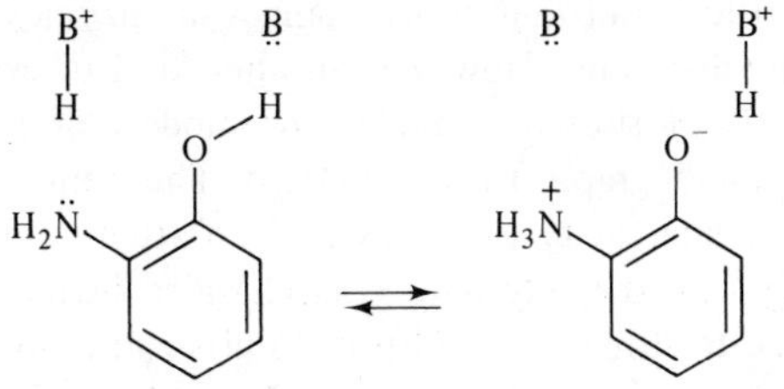

The drive toward charge neutralization in a nonpolar environment suggests that the zwitterion in (1.25) would be very effective. Unfortunately, this tautomer does not exist in benzene. Such charged intermediates in a nonpolar site may well function on an enzyme (1.26).

(1.26) enzyme—CO_2^- H—substrate | B: enzyme—$\overset{+}{N}H_3$

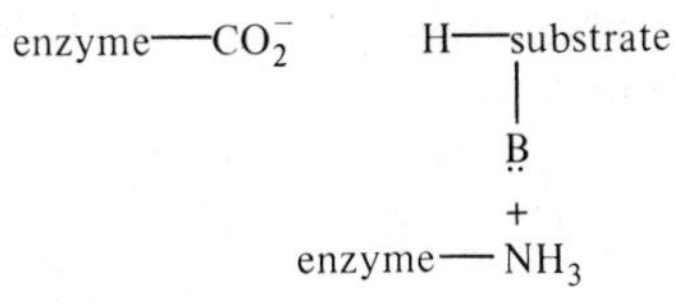

Stabilization of the Transition State

We have looked at various interactions of enzymes with substrates. In a reaction, however, changes in the substrate and in the interaction of substrate with the enzyme must occur as bonding changes. A static picture of enzyme catalysis is insufficient. In the early 1950s, Pauling recognized that enzymatic catalysis would be most effective if the enzyme stabilized the transition state more than it did the products (1.27). This extra stabilization speeds the reaction.

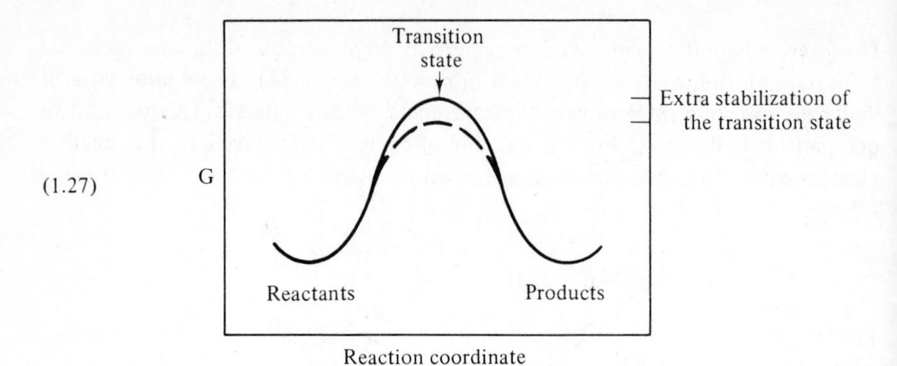

(1.27)

Because the forces of attraction between the enzyme and substrate change during the reaction, the shape of the flexible enzyme would be expected to undergo small changes.

Evidence in support of Pauling's prediction is accumulating. Various probes are being used to study conformational changes in enzymes. Such changes will not be discussed in this book. However, another kind of evidence comes from the study of the fit of substrate models to models of lysozyme (as determined by X-ray crystallography) (Section 1.3). The pentasaccharide can only be fit into the active site cavity of lysozyme by flattening the fourth or D ring. The binding energy of the alternating saccharides increases with additional monomers from one to three, but additional units cause no increase in bonding energy. Apparently, the unfavorable ring distortion balances the favorable bonding interactions of longer chains.

It remains for this distortion to be connected with reaction mechanisms. As seen in (1.28), the reaction can occur through a carbonium-like intermediate. Maximum stabilization of the charge is obtained in the planar structure. The strain of the ring makes it easier to break the carbon–oxygen bond joining rings D and E of the substrate because the oxygen is positioned for maximum overlap with the developing carbonium ion.

(1.28)

R—O Ring D
H^+
$\xrightarrow{-HOR}$

1.5 STEREOSPECIFIC CATALYSIS BY ENZYMES

Most enzymes have more than 100 asymmetric carbons. The enzyme itself also has a chiral three-dimensional shape. Enzymes are remarkably stereoselective reagents. Indeed, most enzymatic reactions are completely stereospecific. This is

(1.29)

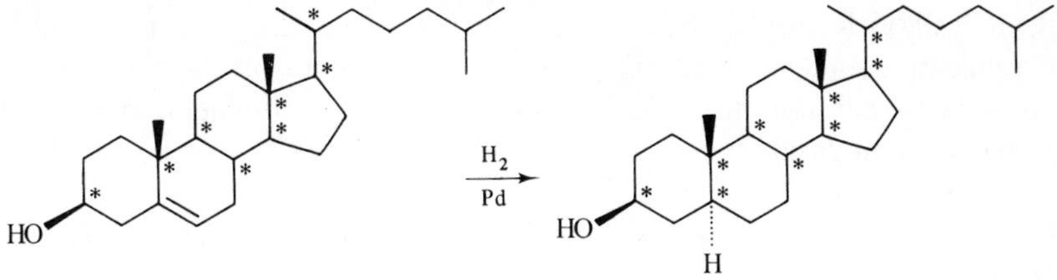

HO → (H_2, Pd) HO, H

not surprising, for the hydrogenation of cholesterol (with eight asymmetric centers) gives a single stereoisomer (1.29).

Enzyme stereospecificity may occur in either the binding step or in a reaction step. Two enzymes catalyze the oxidative deamination of amino acids (1.30). One of these enzymes is specific for L-amino acids; the other, for D-amino acids. (The two enzymes also use different electron acceptors.) Amino acids of the wrong configuration do not inhibit the reaction; thus, they do not bind to the enzyme. In this case the selectivity is in the substrate binding step. Because the enzyme and the amino acids are asymmetric, the amino acids form diastereomeric complexes with a given amino acid oxidase. These diastereomeric states differ so much in energy that only one is populated.

$$H_2O + RCH(\overset{+}{N}H_3)CO_2^- \longrightarrow RC(=O)CO_2^- + 2H + NH_4^+ \tag{1.30}$$

There is another type of enzyme selectivity which is more subtle. The addition of a hydride ion to NAD^+ is stereospecific even though C-4 is not a chiral center in either NAD^+ or NADH (the result of adding a hydride ion to NAD^+). The selectivity can only be detected by the use of isotopes of hydrogen (1.31). Using deuterium labeled substrate, it has been found that hydride ion is incorporated into the *B* side with glyceraldyhyde-3-phosphate dehydrogenase and into the *A* side with lactate dehydrogenase. Several other enzymes that use NAD^+ as an oxidizing agent are stereospecific for one or the other side. One enzyme that is not stereospecific is known.

(1.31)

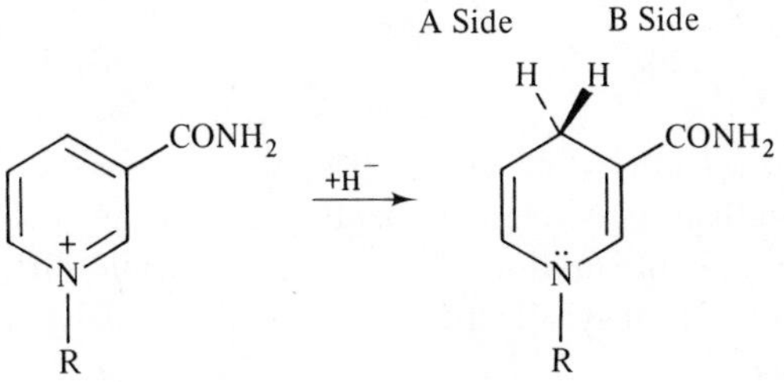

A Side B Side

$CONH_2$, $\overset{+}{N}$, R → ($+H^-$) H H, $CONH_2$, $\ddot{N}$, R

Initially, this stereospecificity at an achiral center was thought to require a three-point attachment by the enzyme (1.32). The hypothetical attachment by the asymmetric enzyme to R, and to the amide group, orients the NAD^+ so that a hydride entering from the *A* side would no longer be enantiomeric to a hydride entering from the *B* side.

(1.32)

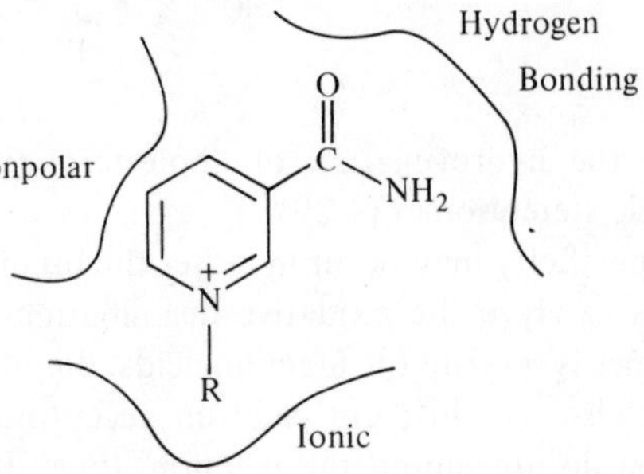

Hydrogen Bonding

Nonpolar

Ionic

While this model accounts for the stereospecific reduction, it is now recognized to be a sufficient but not necessary condition. The faces of NAD^+ are enantiomeric, for the addition of an isotope of hydrogen to give NADH-d generates an asymmetric center at C-4. Transition states for the addition of hydride are enantiomeric in a chemical reduction. However, with an enzyme attached, the transition states are diastereomeric, and the reaction is stereospecific (1.33).

(1.33)

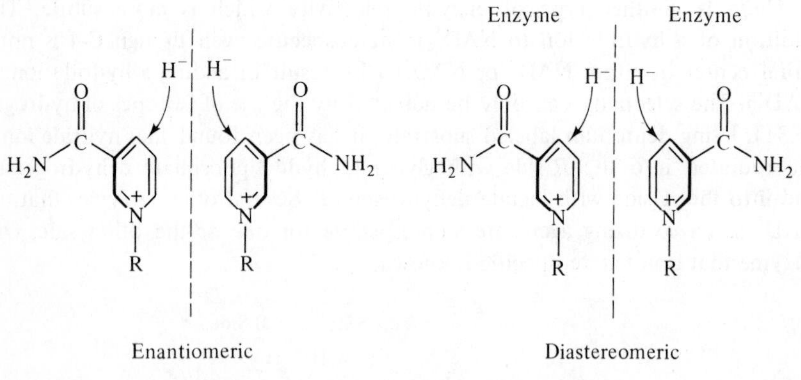

Enzyme Enzyme

Enantiomeric Diastereomeric

The high stereoselectivity of biochemical reactions is necessary. If amino acids of the D-configuration were frequently inserted into polypeptide chains, enzymes would have configurational isomers. These would differ in their shape and activity. The efficiency of the cell's chemical catalysis would be lost.

Stereoselectivity is also necessary for small molecules used as substrates. Because enzymes are asymmetric, they have a preference for one or another substrate configuraiton, yet to use both stereoisomers would require additional enzymes. A larger cell would be required, thus lowering the efficiency of transport processes. Life cannot afford that inefficiency, so biochemical reactions are amazingly stereospecific.

1.6 COENZYMES

Many enzymes catalyze similar reactions. An amino group of a particular substrate may be oxidized to a carbonyl group. The design of an efficient catalyst for such a reaction requires an efficient electron acceptor. This could have evolved in two ways: either each enzyme evolved such a group (acceptor), or a small molecule having this function (as acceptor) might be used by several enzymes. Evolution took the latter course. The class of small molecules used by several enzymes are referred to as *coenzymes.* Some coenzymes carry reducing or oxidizing power between different enzymes; others carry high-energy groups. Still other coenzymes may help to initiate similar reactions on a variety of enzymes. The chemistry of coenzymes should reveal general features of enzyme mechanism. The particular functions of several coenzymes will be discussed in the following chapters.

1.7 SUMMARY

Enzymes are the matrices for biochemical reactions. They bind to substrates and provide a highly structured active site for the reaction. Their catalysis uses the same chemical principles that govern reactions in the chemical laboratory. Because the catalysis may involve a wide variety of coordinated factors, the mechanism of biochemical reactions is potentially a rich literature for the chemist interested in mechanisms.

REFERENCES

Protein Structure

H. B. Bull, *Adv. Enz.,* **1**, 1 (1941).
R. E. Dickerson and I. Geis, *The Structure and Action of Proteins,* Harper & Row, Publishers, New York (1968).
G. Kartha, *Acc. Chem. Res.,* **1**, 374 (1968).
F. M. Richards, *Ann. Rev. Biochem.,* **32**, 269 (1963).

Catalysis by Enzymes

M. L. Bender, *Mechanisms of Homogeneous Catalysis from Protons to Proteins,* Wiley-Interscience, New York (1971).

T. C. Bruice and S. J. Benkovic, *Bioorganic Mechanisms,* Vol. I and II, W. A. Benjamin, Inc., New York (1966).

W. R. Jencks, *Ann. Rev. Biochem.,* **32**, 639(1963).

W. R. Jencks, *Catalysis in Chemistry and Enzymology,* McGraw-Hill Book Co., New York (1969).

J. F. Riordan and M. Sokolovsky, *Acc. Chem. Res.,* **4**, 353 (1971).

I. A. Rose, *Ann. Rev. Biochem.,* **35**, 23 (1966).

Spectra of Nicotinamides

E. M. Kosower, *Biochem. Biophys. Acta,* **56**, 474 (1962).

E. M. Kosower, *Molecular Biochemistry,* McGraw-Hill Book Co., Inc., New York, (1962).

G. Maggiora, H. Johansen, and L. L. Ingraham, *Arch. Biochem. Biophys.,* **131**, 352(1969).

2 HIGH ENERGY COMPOUNDS

2.1 INTRODUCTION

A source of energy is a necessary requirement for any organism. Energy is required by the organism for many functions, including movement, growth, and even thinking.

A living cell is a very highly organized structure, this high organization being typical of life. When it disappears the cell dies. High organization means low entropy, so that a living cell by necessity must have a lower entropy than a dead cell. An important use of free energy is to lower the entropy of the organism, and since the second law of thermodynamics states that entropy may only increase or remain constant, the order required for life must produce disorder elsewhere in the environment.

Cells maintain their order by feeding on compounds rich in free energy and producing compounds high in entropy. Thus, order is consumed and disorder is drained off from the organism. Because small molecules have more entropy per gram than large molecules, and gases have a higher entropy per gram than solids or liquids, the metabolic products, carbon dioxide and water, have a larger entropy per gram than foodstuffs. All of the carbon dioxide and much of the water is excreted as a gas, and most foods are solids or liquids.

The free energy that an organism consumes depends both on the organizational requirements of the cell and the enthalpic requirements of the cell. Whenever the organism needs more energy for organization (to lower the entropy), less free energy is available to the organism for the more conventional uses.

The primary source of energy for biological use is the sun. Photosynthesis in plants transforms radiant energy from the sun into chemical energy for use by the plant or by whatever eats the plant. Photosynthesis is a relatively efficient process, but very little of the sun's total energy that reaches the earth is converted into biological energy. Because of deserts, oceans, and rocky land, only about one-tenth of one percent of the sun's energy is captured by photosynthesis. In a dense forest, about 3% of the sun's energy is utilized.

2.2 CHEMICAL FORMS OF ENERGY

For the efficient storage of chemical energy, a compound must meet three requirements: it must undergo a reaction that produces considerable amounts of free energy, it must be kinetically stable (otherwise the compound will react before it is needed), it must react with a compound that is readily available to the cell (if not, the cell would need to store and transport two special compounds when energy is needed).

When we consider what reactants are readily available to an aerobic cell, we immediately think of oxygen and water. These are two reactants used by the cell to produce energy. Because of the ready availability of oxygen, the cell could store energy as reducing agents. (Since oxygen is a strong oxidizing agent, all reduced compounds are potentially a good source of chemical energy for the cell.) And since water is also readily available, the cell may store chemical energy in compounds that have a high free energy of hydrolysis. Esters—and in particular, anhydrides—can be used for energy. Anhydrides of phosphates and mixed anhydrides of phosphates and carboxylic acids have a high free energy of hydrolysis and are used by the cell to take advantage of the ubiquitous reactant, water.

The kinetic stability of the compounds is an important factor. The compound should not leak energy by reacting in the absence of an enzyme. For instance, acid chlorides that react readily with water would be unsuitable. There must be mechanistic restrictions to the cleavage of the high-energy bonds.

Most enzymatic reactions are coupled.

In a hypothetical example (2.1), the half-reaction (b) may require free energy and the half-reaction (c) may liberate free energy. The overall reaction, (a),

$$
\text{(2.1)} \qquad
\begin{aligned}
&\text{(a)} \quad A + B \underset{}{\overset{\text{enzyme}}{\rightleftharpoons}} C + D \\
&\text{(b)} \quad A \rightleftharpoons C \\
&\text{(c)} \quad B \rightleftharpoons D
\end{aligned}
$$

reflects this balance. For some enzymatic reactions, an equilibrium constant near unity may be desired. (Both the reactants and products may be needed by other enzymes.) For the synthesis of structural materials, it is advantageous to drive reactions toward products so that intermediates do not accumulate. Reactions leading to biopolymers are frequently coupled with the hydrolysis or oxidation of energy-rich compounds.

This trick of driving reactions is also utilized by synthetic organic chemists. For example, mixing a carboxylic acid in aqueous solution with an amine will not give an amide because of the unfavorable equilibrium. However, the highly reactive PCl_5 can be added to the carboxylic acid to first form the acid chloride. This compound can be reacted with an amine to form an amide. One may say that the synthesis of the amide from a carboxylic acid and an amine can be driven by the hydrolysis of PCl_5.

Examples of compounds that are commonly used to store energy in organisms will be presented and discussed in the remainder of this chapter. The list is not complete, but the compounds chosen contain representative structural features of high-energy compounds.

2.3 ATP AND CHARGE REPULSION

The primary chemical energy in the cell is from the hydrolysis of adenosine triphosphate, shown in (2.2). Adenosine triphosphate will be abbreviated hereafter, as is the custom in most biochemical literature, as ATP. ATP acts like electricity in a modern country. Many sources of power–coal, oil, water, and nuclear–are converted into electricity and fed into the transmission lines. Whenever an industry or home needs energy, they draw upon this source of electricity. Similarly, the cell uses many types of reactions to produce ATP, and, whenever energy is needed for motion, synthesis, or any other use, it draws upon this supply of ATP. The cellular concentration of ATP is about 2-15 millimolar.

Notice in (2.2) that ATP has two anhydride bonds. Either of these are susceptible to hydrolysis by water. If the first anhydride bond is hydrolyzed, the products are adenosine monophosphate and pyrophosphate ion. Adenosine monophosphate is commonly abbreviated AMP; the structure is shown in (2.2). If the second phosphate bond is hydrolyzed, the products will be adenosine diphosphate and phosphate ion. The structure of adenosine diphosphate is also given in (2.2) and, in accordance with the previous abbreviations, the initials ADP are used as a designation. The hydrolysis to ADP and phosphate ion at pH 7 has a free-energy of -9.6 kcal. The hydrolysis to AMP and pyrophosphate ion

NH_2 N N N N O CH_2O—P(=O)(O^-)(δ^+)—O—P(=O)(O^-)(δ^+)—O—P(=O)(O^-)(δ^+)—OH

OH OH

Adenosine triphospate (ATP)

(2.2)

NH_2 N N N N O CH_2O—P(=O)(O^-)(δ^+)—O—P(=O)(O^-)(δ^+)—OH

OH OH

Adenosine diphosphate (ADP)

NH_2 N N N N O CH_2O—P(=O)(O^-)(δ^+)—OH

OH OH

Adenosine monophosphate (AMP)

at pH 7 produces almost the same energy, -9.8 kcal. The energy of both of these reactions is utilized by the cell for many functions needing energy. ATP can also hydrolyze to produce AMP and two phosphate ions. This reaction is the sum of the reaction to produce AMP and pyrophosphate ion, -9.8 kcal, and the hydrolysis of pyrophosphate ion to produce two phosphate ions with free energy of -8.6 kcal. The total free energy for the reaction is the sum, -18.4 kcal. Thus ATP has some versatility. It can produce two levels of energy, depending upon how much energy the cell needs. These values are tabulated in Table 2.1.

TABLE 2.1 FREE ENERGIES OF ATP HYDROLYTIC REACTIONS AT pH 7 AND 0.2 IONIC STRENGTH

$ATP + H_2O \rightarrow ADP + P^*$	−9.6	kcal/mole
$ATP + H_2O \rightarrow AMP + PP^*$	−9.8	
$ATP + 2H_2O \rightarrow AMP + 2P^*$	−18.4	

All of the above values for ATP hydrolysis are in the absence of magnesium ion. Magnesium ion will chelate with all these species, and to a different extent with each. Therefore, magnesium ion will change the equilibrium and produce a new value for the free energy change. In 10^{-3} *M* magnesium ion and at an ionic strength of 0.2, the free energy for the reaction of ATP to produce ADP and phsophate ion is −8.8 kcal. This may be a more realistic value to use for calculations of in vivo reactions than the value of −9.6 kcal that was measured in the absence of magnesium ions.

The instability of ATP towards hydrolysis is understandable when we consider the repulsions between the negative charges of the oxygen off the main P–O–P chain. Oxygen is much more electronegative than phosphorous so that a phosphorous-oxygen double bond is primarily a semipolar bond with a high negative charge on the oxygen not a part of the main P–O–P chain (2.3). The repulsion between these oxygens is an important factor in adding to the instability of ATP.

$$R-O-\overset{\overset{\displaystyle O}{\|}}{\underset{\underset{\displaystyle O^-}{|}}{P}}-OH \qquad\qquad R-O-\overset{\overset{\displaystyle O^-}{|}}{\underset{\underset{\displaystyle O^-}{|}}{P^+}}-OH \tag{2.3}$$

Expanded valence shell — Coordinate covalent bond

Another electrostatic factor adds to the thermodynamic instability of ATP. The positively charged phosphorous atoms in the semipolar double bond tend to drain electrons off the bridging oxygens by means of both electrostatic attraction and *d* orbital expansions by the phosphorous atoms. These effects are not complete. Calculations predict that the bridging oxygens are still slightly negative, but the repulsion between the positively charged phosphorous atoms is only slightly compensated for by the weakly negative oxygens.

The above electrostatic interactions are affected by the number of phosphate groups and the protonation of the phosphate groups. The effect of the number of phosphate groups is fairly clear. The more groups that are attached together, the greater are the charge repulsions, and the more negative is the free energy of hydrolysis. The affect of protonation on the free energy of hydrolysis is more difficult to assess. One of the greatest problems is in unravelling its affect on electrostatic repulsions versus free energies involved in neutralization reactions. The free energy of neutralization can be quite large. The affect of protonation on the free energy of hydrolysis depends upon whether the products or the reactants are stronger bases. This in turn determines whether protons are used or produced in the reaction.

When protons are produced in a reaction, there are two effects. The consumption of protons is exactly the reverse of the effects that occur when protons are produced. When protons are produced at pH 7 they are partially neutralized. This makes the enthalpy more negative and therefore the free energy more negative. Protons are a small positive charge and because of this they are highly solvated. This causes the entropy to decrease and the free energy to become more positive. However, this effect is less than the enthalpy of neutralization.

The overall effect of producing protons at pH 7 is to make the free energy of the reaction more negative. These predictions are born out by experimentation. No protons are produced when ATP hydrolyzes to ADP and phsophate ion at pH 5. The free energy of hydrolysis is -8.2 kcal and the entropy increase is +9.0 entropy units. At pH 7 the hydrolysis produces one proton and the entropy becomes negative (-17.0 eu) from solvation, but this change is more than compensated for by a more negative enthalpy, making a more negative free energy (-9.2 kcal) than at pH 5. This trend continues. The higher the pH, the more negative is the entropy for the hydrolysis of ATP to ADP and phosphate ion.

The other effect of protonation has been neglected in the above discussion. This is the affect of protonation on the intrinsic instability of ATP. How does protonation affect electrostatic interactions? Protonation would tend to neutralize the repulsions between the negatively charged oxygens, but it would tend to increase the repulsion between the positively charged phosphorus atoms. The easiest way to test the effect of protonation is to study how the intrinsic stability depends upon esterification. Esterification neutralizes the charges (as does a proton) but eliminates the problem of producing a proton in solution. Experimental results with esterification have shown that the latter effect, the effect of increasing the repulsion between the positively charged phosphorus atoms is more important. The free energy of hydrolysis is more negative for ADP than for pyrophosphate ion.

Although the hydrolysis of anhydride bonds of ATP is highly exothermic, ATP fills the requirements for a biological source of energy. It can store energy because the high-energy bonds are not hydrolyzed in the absence of specific enzymes. For hydrolysis to occur, the oxygen of water or of hydroxide ion must attack a positive phosphorus. The high negative charge of the oxygens of ATP strongly repels the nucleophiles, preventing hydrolysis. Enzymes that cleave ATP require a divalent cation to function. The cation partially neutralizes the excess negative charge, permitting a nucleophile to attack the phosphorus.

2.4 OTHER HIGH-ENERGY COMPOUNDS—THE IMPORTANCE OF RESONANCE

There are several compounds other than ATP that store energy by virtue of a high free energy of hydrolysis. These compounds function as reactive intermediates for temporary storage of energy. Some are used in the conversion of energy to ATP; others are used in the conversion of ATP energy to some other use in the cell. A few of these compounds will be discussed to identify various structural features that can be associated with a high energy of hydrolysis. For these compounds, as well as for ATP and ADP, acid or base catalysis is required for hydrolysis to occur.

Acetyl phosphate, the anhydride of acetic and phosphoric acids (2.4) has a free energy of hydrolysis of −10.10 kcal at pH 7.0. The free energy of hydrolysis has this large negative value because the product of the reaction, acetate ion, is stabilized by resonance, but the reactants have little stabilization by resonance.

$$CH_3C(=O){-}O{-}P^{\delta+}(=O)(OH){-}O^- \xrightarrow{H_2O} CH_3C(=O){-}O^- + O{=}P(O^-)(OH){-}OH \tag{2.4}$$

$$\updownarrow\!\!\!\!\times \qquad\qquad\qquad \updownarrow$$

$$CH_3C(-O^-){=}O^+{-}P^{\delta+}(=O)(OH){-}O^- \qquad\qquad CH_3C(=O^-)... O \;\; (CH_3C(-O)=O^-)$$

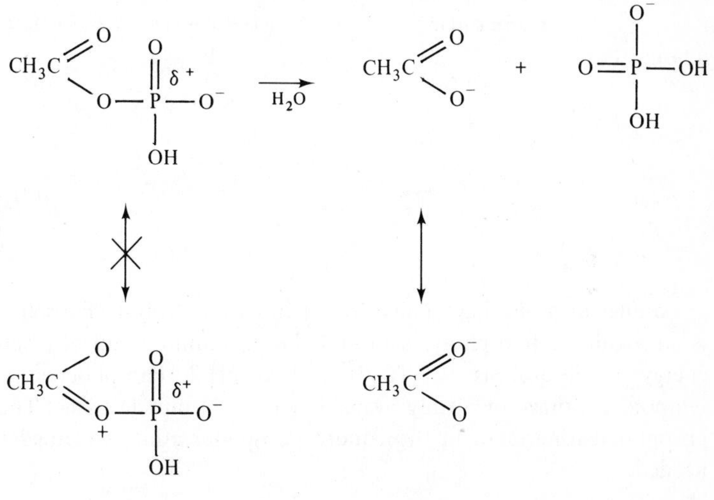

Resonance in the reactant, acetyl phosphate, places a positive charge adjacent to the already positive phosphorus atom, thereby causing resonance in the reactant to be of minor consideration. The large resonance stabilization of the products and small stabilization of the reactants causes a relatively large free energy of hydrolysis.

Again, 1,3-diphosphoglyceric acid, an anhydride of phosphoric and 3-phosphoglyceric acid (2.5), has a large free energy of hydrolysis for exactly the same reasons as does acetyl phosphate. The free energy of hydrolysis is slightly larger than that of ATP at pH 7—−11.80 kcal—probably because of repulsions between the negative charges of the two phosphate groups in the reactant.

(2.5)

$$\mathrm{HO\overset{O}{\overset{\|}{P}}(O^-)OCH_2CHOHC}\!\left\langle\begin{matrix}=O\\ O-\overset{O}{\overset{\|}{P}}{}^{\delta+}(O^-)-OH\end{matrix}\right. \;\leftrightarrow\!\!\!\!\!\times\; \mathrm{HO\overset{O}{\overset{\|}{P}}(O^-)OCH_2CHOHC}\!\left\langle\begin{matrix}-O^-\\ =\overset{+}{O}-\overset{O}{\overset{\|}{P}}{}^{\delta+}(O^-)-OH\end{matrix}\right.$$

Phosphoenol pyruvate (2.6) has an even larger free energy of hydrolysis, −14.80 kcal at pH 7. The product is pyruvic acid in the keto form. Most of the driving force of this reaction is the result of the large free energy difference between keto and enol forms of pyruvic acid. In general, ketones are much more stable than their enolic tautomers.

(2.6)

$$\mathrm{CH_2{=}C(OPO_2H)CO_2^-} \xrightarrow{H_2O} \mathrm{HO-\overset{O}{\overset{\|}{P}}(O^-)-OH} + \mathrm{CH_2{=}C(OH)CO_2^-} \longrightarrow \mathrm{CH_3\overset{O}{\overset{\|}{C}}CO_2^-}$$

Amides may also have a high free energy of hydrolysis. Phosphocreatine (2.7) is an amide of phosphoric acid and the guanidino group of creatine. The free energy of hydrolysis is −10.30 kcal at pH 7. Phosphocreatine provides a temporary storage of readily available energy in muscle tissue. The muscle uses phosphocreatine, and, in turn, more phosphocreatine is formed from ATP as needed.

(2.7)

$$\mathrm{HO-\overset{O}{\overset{\|}{P}}(O^-)-NH\overset{NH}{\overset{\|}{C}}N(CH_3)CH_2COOH}$$

Guanidine is a strong base because its conjugate acid is highly stabilized by resonance. The positive charge may be equally shared by all three nitrogens in the protonated form (2.8). A phosphoguanidine, however, is not a strong base. One of the resonance forms of the conjugate acid has been eliminated (2.9).

(2.8)

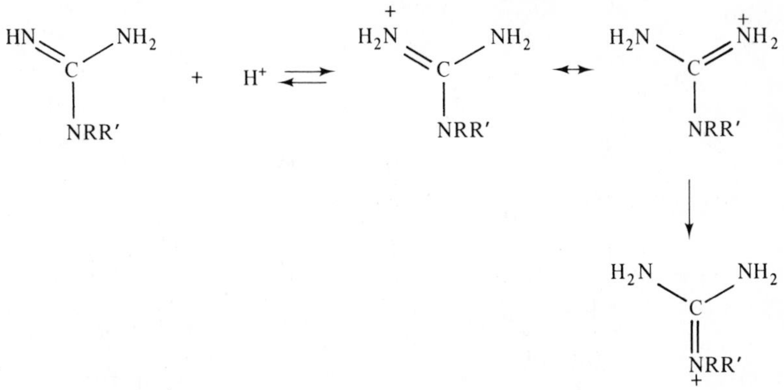

$$HN{=}C(NH_2)(NRR') + H^+ \rightleftharpoons H_2\overset{+}{N}{=}C(NH_2)(NRR') \leftrightarrow H_2N{-}C({=}\overset{+}{N}H_2)(NRR') \rightarrow (H_2N)(NH_2)C{=}\overset{+}{N}RR'$$

The nitrogen that is phosphorylated cannot be positive since this would place a positive charge adjacent to the positive phosphorus. However, on hydrolysis the phosphoguanidine would become a strong base again and accept a proton to form a resonance-stabilized guanidinium ion. Thus a partially stabilized reactant hydrolyzes and is protonated to produce a highly stabilized cation with the release of a large free energy of hydrolysis.

(2.9)

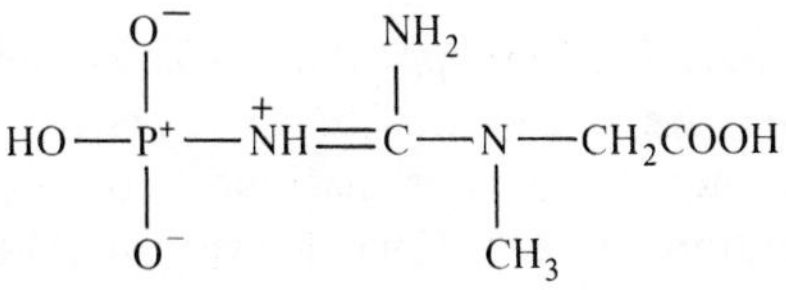

$$HO{-}P^+(O^-)(O^-){-}\overset{+}{N}H{=}C(NH_2){-}N(CH_3){-}CH_2COOH$$

2.5 NADH AND THE IMPORTANCE OF HETEROAROMATIC COMPOUNDS

Because of the ready availability of oxygen to an aerobic cell, any reducing agent, and in particular any organic compound, is a source of energy to the cell. These include carbohydrates, fats, and proteins–all foods for the organism. Simple, fast mechanisms to oxidize these compounds are not always available to the cell. The cell oxidizes all foodstuffs mentioned above by complex pathways. In so doing it stores the reducing power as dihydrodiphosphopyridine nucleotide or dihydrotriphosphopyridine nucleotide (2.10). These two compounds are symbolized by NADH and NADPH respectively to represent nicotinamide adenine dinucleotide (reduced) and nicotinamide adenine dinucleotide phosphate (extra phosphate and reduced). The reducing power can be used for energy or for synthesis of organic compounds.

If organisms merely converted reducing power to heat and acted as heat machines instead of chemical plants, then mechanism and storage of reducing power would be of no consequence. Uncontrolled burning of fuels would

(2.10)

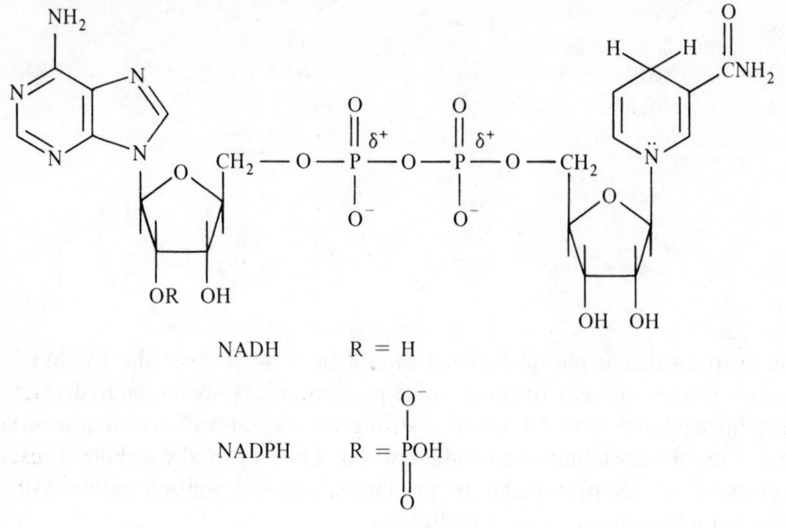

NADH R = H

NADPH R = $\overset{\text{O}^-}{\overset{|}{\text{P}}}\text{OH}$ (with =O)

produce heat for power. This is impossible for an organism because of its low operating temperature. An organism with an operating temperature of 37° C (310° Kelvin) could not be a heat machine because the efficiency for an environmental temperature at 25° C (298° Kelvin) would be less than 4%.

$$\text{Efficiency} = \frac{310^\circ - 298^\circ}{310^\circ} \times 100 = 3.8\%$$

Both compounds used for storage of reducing power–NADH and NADPH–are reduced pyridine compounds. Both are two-electron-hydride donating agents with a potential of −0.320 volts. The hydride ion is donated readily to produce a stable pyridinium ring. The oxidized forms of the compounds are also called diphosphopyridine nucleotide and triphosphopyridine nucleotide. The structures of these compounds are shown in (2.11). The symbols used to designate the oxidized forms are NAD^+ and $NADP^+$.

(2.11) [reduced pyridine, R] → [pyridinium, R] + H^-

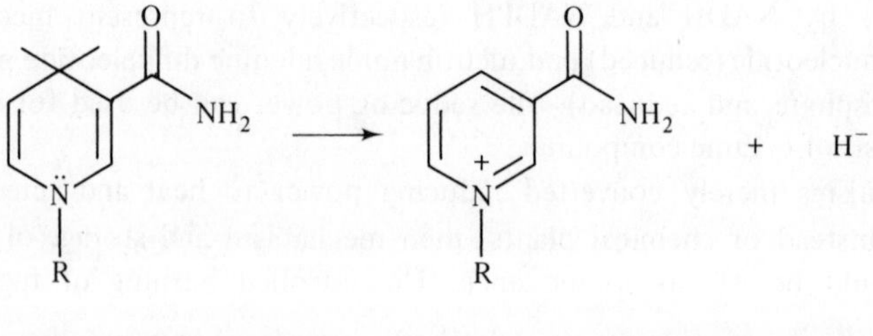

The reducing power of pyridine nucleotides is used both for organic synthesis and for energy. Organic redox pairs are severely limited in number. The reduction by hydrogen of most double bonds is strongly exothermic. In contrast, aromatic rings are reduced with difficulty compared to olefins, because reduction is accompanied by the loss of resonance energy. Another difficulty for many organic oxidations and reductions is the high activation energies for reaction. The pyridine nucleotides are a useful redox pair because they react rapidly under mild conditions.

The pyridine nucleotides are used by the cell as a source of hydride ion or as an agent to remove a hydride ion. A typical reduction is the reduction of a carbonyl group to an alcohol. Thus NADH and NADPH are the cell's reagents that correspond to the sodium borohydride or lithium aluminum hydride used by an organic chemist. The positive charge on the pyridinium ring enables the pyridine nucleotides to remove a hydride ion from a substrate. The increased resonance of the pyridinium ring over the dihydropyridine ring appears to make the reaction highly reversible. The pyridine nucleotides are used by the cell for two electron oxidation-reductions whereas the flavins, which are discussed in Chapter 5, are used for both one and two electron oxidation-reductions.

The oxidation by molecular oxygen to produce energy converts the reducing power of the reduced pyridine nucleotides to ATP. This occurs in a series of complex reactions called *the electron transport chain.* Essentially, electrons are transferred from the reduced pyridine nucleotide to oxygen, and in the process ADP is esterified with phosphate ion to produce ATP. The actual mechanism of how this is done is still unknown. The reduced pyridine nucleotides are in concentrations too low to act for any substantial storage of energy. They rapidly alternate between the oxidized and reduced states to facilitate energy conversions. The great storage of reducing power of an organism is in the form of fats or glycogen.

2.6 SUMMARY

Chemical energy in living organisms is stored in compounds that are hydrolyzed or oxidized to liberate energy. The energy can be used to drive other reactions in coupled reactions. Charge repulsion and resonance energy are concepts that can be used to describe the structures of reactants and products of high-energy reactions.

REFERENCES

NADH Reactions

R. H. Abeles, R. F. Hutton, and F. H. Westheimer, *J. Am. Chem. Soc.*, **79**, 712 (1957).

S. Chaykin, *Ann. Rev. Biochem.*, **36**, 149 (1967).
D. Mauzerall and F. H. Westheimer, *J. Am. Chem. Soc.*, **77**, 2261 (1955).
F. Schlenk, *Adv. Enz.*, **5**, 207 (1945).
T. P. Singer and E. B. Kearney, *Adv. Enz.*, **15**, 79 (1954).

Phosphates

F. Lipmann, *Adv. Enz.*, **1**, 99 (1941).

3 PYRIDOXAL AND SCHIFF BASES: TAUTOMERIC SHIFTS AND ELIMINATIONS

3.1 INTRODUCTION

To the organic chemist, conversions (3.1)-(3.5) look very different. These reactions can be described as a transamination (3.1), as a rearrangement (3.2), as an elimination and rehydration (3.3), as a decarboxylation (3.4), and as a reverse condensation (3.5). For the chemist, each conversion would require a separate strategy and different reagents.

$$RC(=O)CO_2^- + R'CH(\overset{+}{N}H_3)CO_2^- \rightleftharpoons RCH(\overset{+}{N}H_3)CO_2^- + R'C(=O)CO_2^- \quad (3.1)$$

$$\underset{\text{Serine}}{HOCH_2CH(\overset{+}{N}H_3)CO_2^-} \longrightarrow \underset{\text{Pyruvate}}{CH_3C(=O)CO_2^-} \quad (3.2)$$

(3.3)

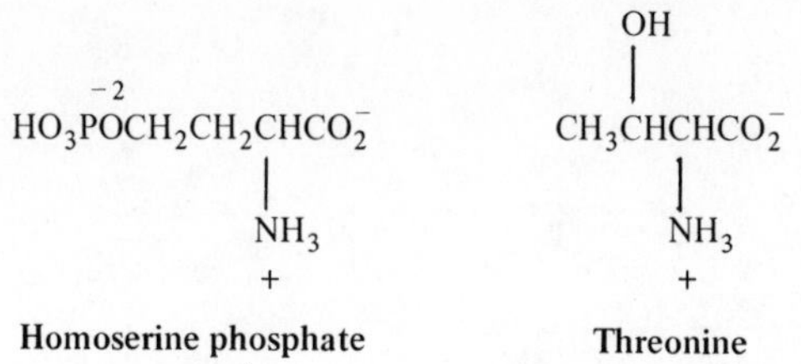

$$\underset{\overset{|}{\overset{+}{NH_3}}}{HO_3\overset{-2}{P}OCH_2CH_2CHCO_2^-} \qquad CH_3\overset{OH}{\overset{|}{C}}HCHCO_2^- \text{ with } NH_3^+$$

Homoserine phosphate **Threonine**

(3.4)

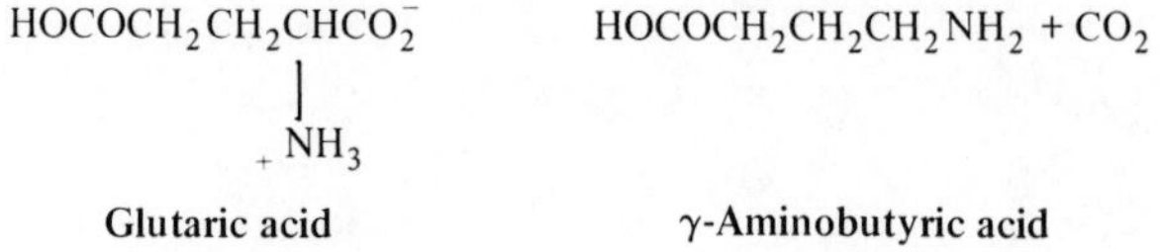

$$HOCOCH_2CH_2CHCO_2^- \,(\overset{+}{N}H_3) \qquad HOCOCH_2CH_2CH_2NH_2 + CO_2$$

Glutaric acid **γ-Aminobutyric acid**

(3.5)

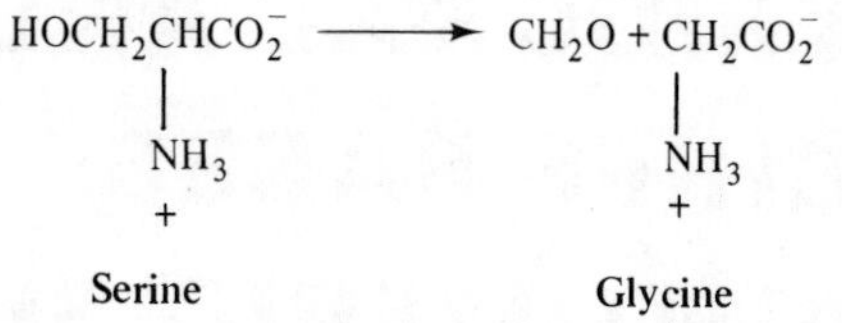

$$HOCH_2CHCO_2^- \,(\overset{+}{N}H_3) \longrightarrow CH_2O + CH_2CO_2^- \,(\overset{+}{N}H_3)$$

Serine **Glycine**

The common feature of these biochemical reactions and of many mechanistically similar reactions of amino acids is the requirement for pyridoxal or coenzyme B_6. The term *coenzyme* or *cofactor* refers to biochemical function; the term *vitamin,* to dietary requirement. Three structures for vitamin B_6 commonly occur. These are named pyridoxal (3.6a), pyridoxine (3.6b), and pyridoxamine (3.6c). The lower oxidation states of the vitamins (3.6b and c) are readily converted to the active coenzyme in the body. The active form of B_6 has been identified by adding the individual compounds to purified enzyme systems and assaying for activity.

(3.6)

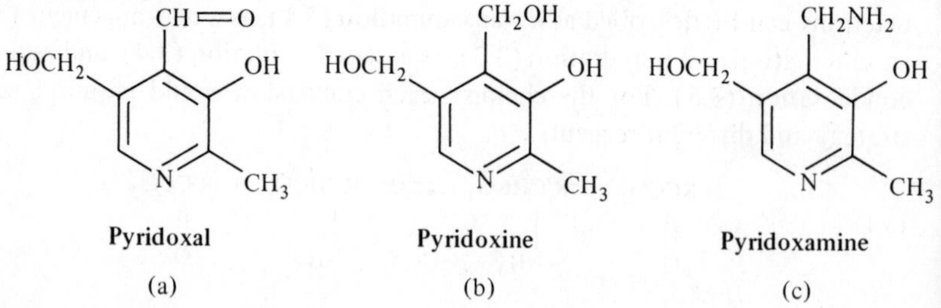

CH=O, $HOCH_2$, OH, N, CH_3 — **Pyridoxal** (a)

CH_2OH, $HOCH_2$, OH, N, CH_3 — **Pyridoxine** (b)

CH_2NH_2, $HOCH_2$, OH, N, CH_3 — **Pyridoxamine** (c)

Evidence indicates that the active form of pyridoxal is an imine, or Schiff base (3.7a). Enzyme-bound pyridoxal can be separated. However, if the enzyme-pyridoxal complex is treated with sodium borohydride (a reagent known to reduce Schiff bases to amines), the product is not pyridoxine (3.6b).

(3.7)

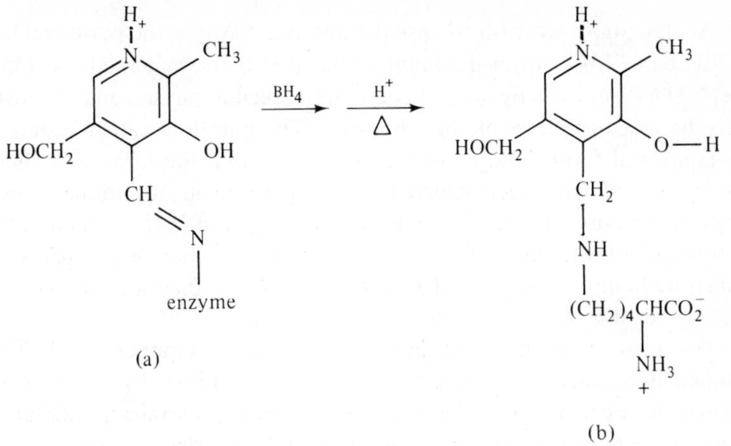

(a) $\xrightarrow{BH_4}$ $\xrightarrow[\Delta]{H^+}$ (b)

Pyridoxal can no longer be separated from the enzyme. Acid hydrolysis of the enzyme followed by separation of the amino acids yields a substituted lysine (3.7b) that was not present in the hydrolysate of the original enzyme.

The imine linkage to the enzyme is first cleaved then later reformed when an enzyme converts an amino acid substrate to products. Pyridoxal is transferred from the enzyme-imine to a substrate-imine (3.8). When the product is released, the enzyme bound Schiff base is reformed. This change may be described as a *transiminization.*

(3.8)

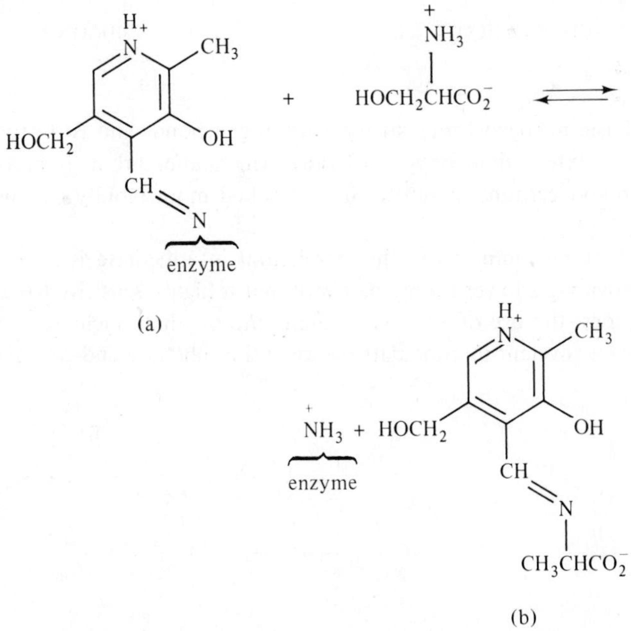

(a) + $HOCH_2CH(\overset{+}{N}H_3)CO_2^-$ $\rightleftharpoons$ $\underbrace{\overset{+}{N}H_3}_{\text{enzyme}}$ + (b)

At this point, we wish to raise the question "Why is the pyridoxal bound as a Schiff base?" Why form the imine (3.8a) if it is to react and be lost in the next step? Questions of why are not commonly asked in physics and chemistry. They may be unavoidable in biology, however. The questions can be restated in a less metaphysical form. Still, in asking, we assume that the forms and chemistry of living systems have been selected by competition in evolutionary history. An enzyme functions to speed reactions. Our original question becomes "Does the formation of the enzyme-bound imine provide a more rapid pathway to the substrate-bound imine?" If the answer is yes, we then seek to find structural reasons for this rate difference.

The relative reactivity of pyridoxal and a corresponding Schiff base was studied in a reaction forming the semicarbazone (3.9). The Schiff base (3.9a), formed between valine and pyridoxal, reacted with semicarbazide at a rate 50 times as great as that of pyridoxal. The greater reactivity of imines compared to the corresponding aldehydes and ketones is due to the greater basicity of imines.

(3.9)

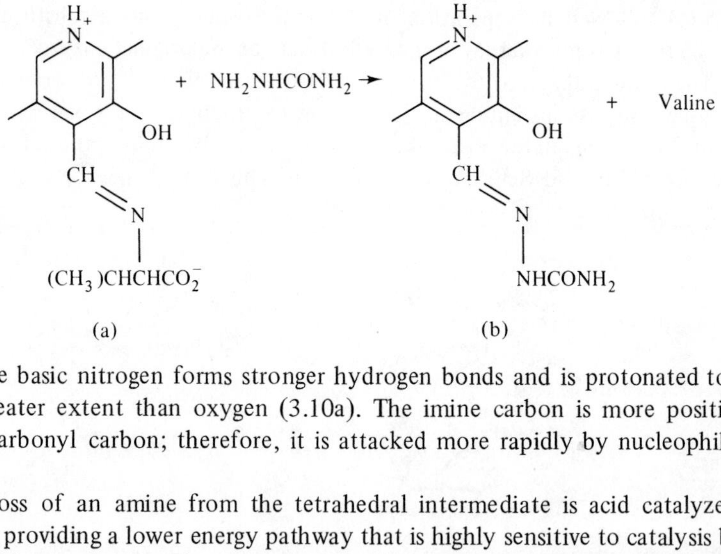

H_+ N … OH … CH=N–$(CH_3)CHCHCO_2^-$ + $NH_2NHCONH_2$ → H_+ N … OH … CH=N–$NHCONH_2$ + Valine

(a) (b)

The more basic nitrogen forms stronger hydrogen bonds and is protonated to a much greater extent than oxygen (3.10a). The imine carbon is more positive than a carbonyl carbon; therefore, it is attacked more rapidly by nucleophiles (3.10b).

The loss of an amine from the tetrahedral intermediate is acid catalyzed. Thus, by providing a lower energy pathway that is highly sensitive to catalysis by acids and bases, the use of an enzyme-imine intermediate facilitates more rapid formation of a covalent intermediate between the substrate and coenzyme.

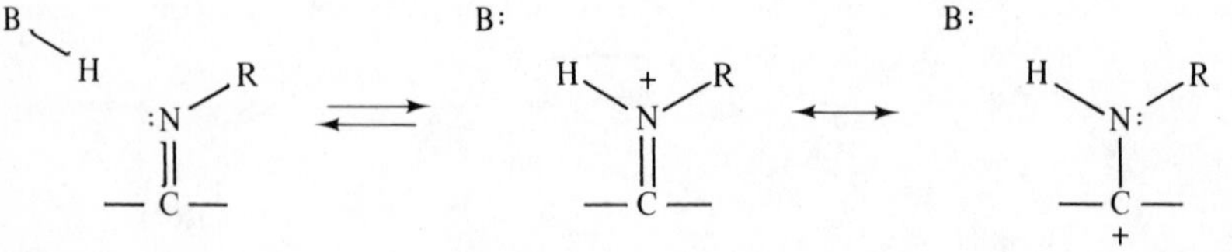

B–H :N(R)=C< ⇌ B: H–N⁺(R)=C< ⟷ B: H–N:(R)–C⁺<

(3.10a)

$$\begin{array}{c} H \quad R \\ \diagdown \ \diagup \\ N \\ \| \\ -C- \end{array} \quad + \quad :Nu \quad \rightleftharpoons \quad \begin{array}{c} H \quad R \\ \diagdown \ \diagup \\ N: \\ | \\ -C- \\ | \\ Nu \end{array}$$

(3.10b)

$$\underset{-H}{\overset{H^+}{\rightleftharpoons}} \quad \begin{array}{c} H \\ | \\ H-N^+-R \\ | \\ -C- \\ | \\ Nu \end{array} \quad \longrightarrow \quad \begin{array}{c} -C- \\ \| \\ Nu \end{array} \quad + \ R\ddot{N}H_2$$

3.2 TRANSAMINATION REACTIONS AND THE STRUCTURE OF REACTION INTERMEDIATES

We can now begin to investigate the structural features of the substrate-coenzyme Schiff base that are responsible for the versatility of the coenzyme. Since the transamination reaction (3.1) proceeds either with or without an enzyme, it is a particularly useful reaction to study. Structurally modified pyridoxals can be tested for catalytic activity in the absence of an enzyme. The proposed mechanism of transamination is given in (3.11). Reversal of this pathway with an α-keto acid followed by cleavage of the imine (3.11a) would form an amino acid.

Let us now identify individual features of this mechanism and examine evidence for each. First, the active form of the Schiff base is a pyridinium compound [(3.11a) rather than (3.9a)]. The positive charge on the ring is essential for catalysis. Pyridoxal itself has pK_a's of 4.2 and 8.7. That at 4.2 corresponds to the ionization of the phenolic hydroxyl; that at 8.7, to the ionization of the pyridinium ring. These pK_a's are very different than those of phenol–9.9–and of pyridinium hydrochloride–5.4. We will discuss the experimental basis used to assign the reversal in order; then we will present the structural basis for this difference in acidity.

The basis of the assignment was the shift of the maximum that accompanies the titration of pyridoxal with strong acid and with strong base. The removal of a proton from a phenolic OH group is accompanied by a shift to longer wavelength. In contrast, pyridinium salts absorb at nearly the same wavelength

$RCHCO_2^-$ $\underset{H^+}{\overset{-H^+}{\rightleftharpoons}}$ $R\ddot{C}CO_2^-$

(a) (b)

(3.11)

$RCCO_2^-$ $\longleftrightarrow$ $RCCO_2^-$

(c)

$\underset{-H^+}{\overset{H^+}{\rightleftharpoons}}$ $RCCO_2^-$ $\underset{-H_2O}{\overset{H_2O}{\rightleftharpoons}}$ CH_2NH_2 + $RCCO_2^-$

(d)

as the corresponding pyridines. From the large shift in the UV spectra accompanying the removal of the first acidic proton from pyridoxal, it was established that this ionization corresponded to (3.12). Had (3.13) occurred, the large shift to longer wavelength should have occurred when the second acidic proton was titrated.

(3.12)

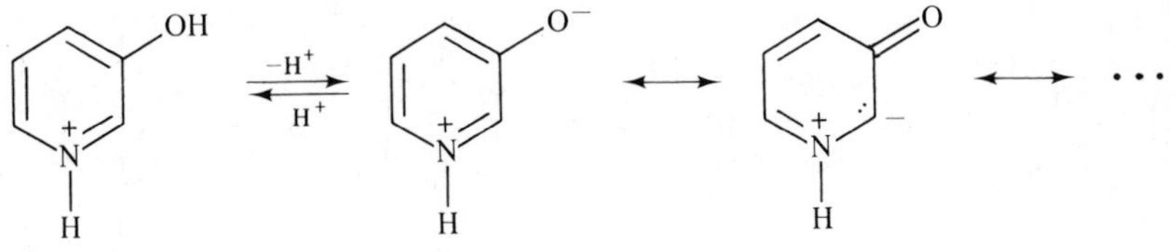

(3.13)

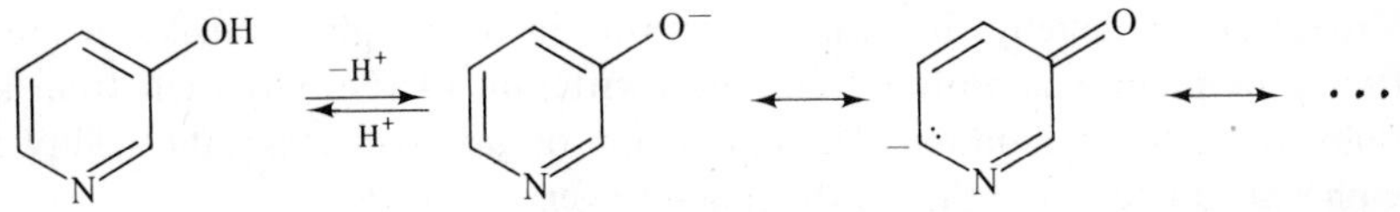

The structural argument for the greatly increased acidity of pyridoxal relative to phenol will be presented in detail. The acid dissociation constant (K_a) for a weak acid reflects the difference in free energies between the acid and its conjugate base. (A hydronium ion is the second product in each reaction.) Factors that destabilize the acid or stabilize the conjugate base increase the acidity of the acid. We wish to compare the ionization of phenol (3.14) to that of pyridoxal (3.15). The charge on the pyridinium ring acts by a field effect to reduce the electron density on oxygen. Also, structure (3.15b) makes a larger contribution to the hybrid structure of pyridoxal than (3.14b) does to phenol. This causes the proton of pyridoxal to ionize more easily than that of phenol.

(3.14)

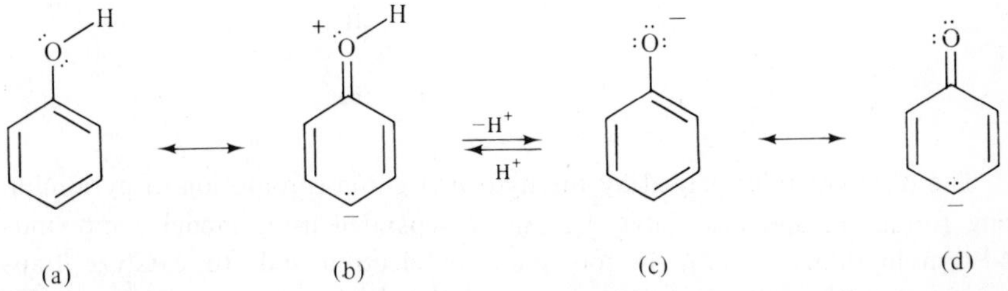

(a) (b) (c) (d)

Because of the charge on nitrogen, the contribution of keto forms such as (3.15d) to the conjugate base of pyridoxal greatly delocalizes the negative charge. This stabilization lowers the energy of (3.15c ↔ d) relative to (3.14c ↔ d). Viewed another way, less work against electrostatic forces is required to remove a proton from the pyridoxal cation than from phenol. The effect of the protonation of the pyridine ring, then, is to lower the energy of the conjugate base of pyridoxal relative to the acid, increasing the acidity of the

(3.15)

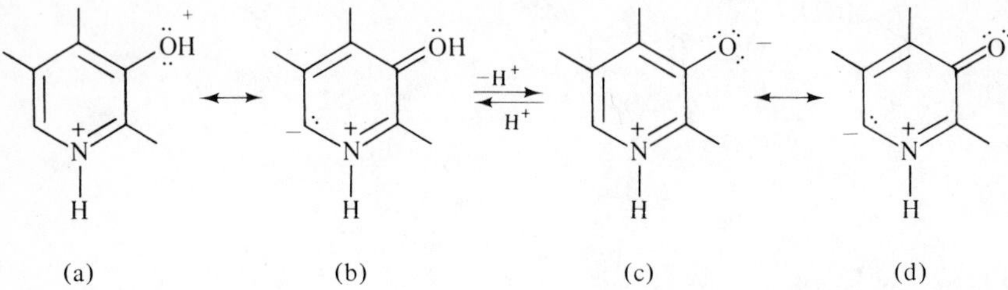

(a) (b) (c) (d)

hydroxyl group thereby. In turn, the ionization of the phenol makes it more difficult to remove a proton from the zwitterion (3.15c ↔ d) than from an unsubstituted pyridinium ion. The interaction of groups increases the acidity of the phenol and decreases that of the ring nitrogen.

A second role of the 3-hydroxyl group of pyridoxal is that of acid catalysis. From the ultraviolet spectra of pyridoxal and 3-methoxylpyridoxal imines, it has been deduced that pyridoxal exists in neutral solution as a mixture of the hydrogen bonded structure (3.16a) and a tautomer in which a proton has been transferred to the imine nitrogen (3.16b).

(3.16)

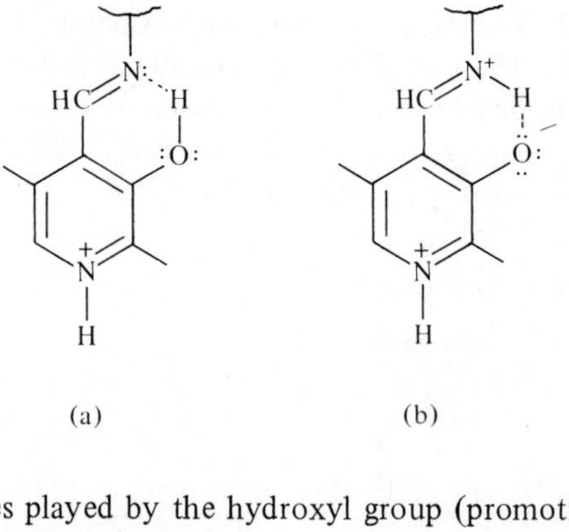

(a) (b)

The different roles played by the hydroxyl group (promotion of pyridinium ring formation and acid catalysis) can be separated using model compounds. 4-Formylpyridine (3.17a) is too poor an electron sink to catalyze transaminations. When this compound is methylated, the resulting cation (3.17b) catalyzes transaminations but at a slower rate than pyridoxal. The positive ring will promote the necessary tautomeric shift discussed below, but the additional acid catalysis of the hydroxyl speeds Schiff base formation and hydrolysis.

In (3.11), one function of pyridoxal is to increase the acidity of the alpha hydrogen of the amino acid. Pyridoxal catalyzes exchange of this hydrogen with D_2O and the racemization of L-amino acids. Both observations are in accord with

(3.17)

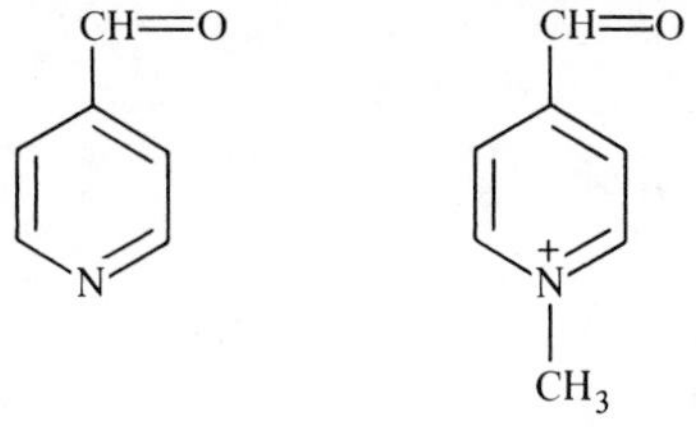

CH═O CH═O

N $\overset{+}{N}$

CH_3

(a) (b)

the planar structure (3.11b ↔ c). Enzyme catalyzed transaminations feature stereospecific protonation to re-form the asymmetric center of the intermediate.

From α-deuterolabeled amino acids, partially deuterated transaminated products may be obtained. Removal of the alpha hydrogen is easier in the imine than in a ketone due to stabilization of the conjugate base (3.11b ↔ c) by resonance and charge neutralization. An objection to the canonical structure (3.11b) may be made because of the adjacent negative charges. These charges may be partially neutralized by the association of the carboxylate group with a cation, such as Ca^{++} or Mg^{++}.

The transamination proceeds by protonation of the intermediate, (3.11b ↔ c), to form (3.11d), the Schiff base of pyridoxamine and an α-keto acid. Hydrolysis of this imine results in the conversion of an amino acid to an α-keto acid. The pyridoxamine can react with another α-keto acid by the reverse series of reactions to form an amino acid and pyridoxal. The oxidation of one amino acid and the reductive amination of a keto acid result from Schiff base formation and proton shifts in the imines.

3.3 ELIMINATION REACTIONS INVOLVING PYRIDOXAL

At this point we are ready to examine the chemistry of the coenzyme substrate intermediates in the enzymatic transformations (3.2) through (3.5). Our approach will be one of rationalizing product formation. Using chemical reasoning, a scheme (paper mechanism) will be proposed. A given mechanism can then be used to predict certain experimental consequences. That these consequences are observed is evidence in accord with the mechanism. Results contrary to prediction disprove a mechanism. Predicted results can only support a mechanism; they are not proof. The mechanism may yet be shown to be incorrect. With this disclaimer, we will seek to understand the reactions [(3.2) through (3.5)] that illustrate additional characteristic biochemical roles of pyridoxal.

The conversion of serine to pyruvate proceeds through the same initial intermediate as transamination. The pyridinium ring of pyridoxal again assists in

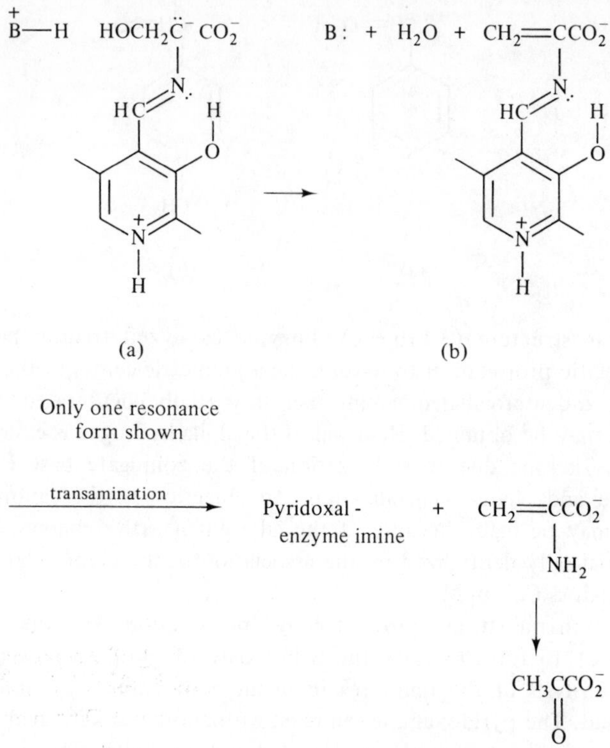

$$\overset{+}{B}—H \quad HOCH_2\ddot{C}^- CO_2^- \qquad \longrightarrow \qquad B: + H_2O + CH_2{=}CCO_2^-$$

(3.18) (a) (b)

Only one resonance form shown

$$\xrightarrow{\text{transamination}} \text{Pyridoxal - enzyme imine} + CH_2{=}C(NH_2)CO_2^- \longrightarrow CH_3C(=O)CO_2^-$$

the removal of the alpha hydrogen, giving (3.18a). The acid catalyzed loss of hydroxide ion from (3.18a) yields the enol of pyruvate imine (3.18b) which then ketonizes and hydrolyzes.

Pyridoxal can also facilitate eliminations from the gamma carbon. Reaction (3.3) is proposed to occur by an elimination of phosphate from homoserine. The complete mechanism is given in (3.19). At this point it is important to reaffirm that this is a plausible scheme. Certain of the details are not proved and perhaps are unprovable. For instance, the intermolecular acid-base reactions [(3.19a → b), (3.19d → e)] are logical reactions, but they have not been proven. Acid and base groups on the enzyme may be responsible for protonating and deprotonating the amine. These acid and base shifts are important, however, in modifying the reactivity of the beta carbon.

That pyridoxal can help stabilize a negative charge on the beta-carbon is the central argument of the mechanism. The tautomeric shifts provide a rationalization for this carbanion character; the elimination of phosphate is a consequence of it. Reversal of step (3.19b → c) and cleavage of the imine provides a path for deuterium exchange. This exchange is observed. Independent evidence for carbanion character at the beta carbon is provided by a trapping experiment. In this experiment the enzyme substrate is cystathionine (3.20a). The intermediate

(3.19)

(a)

(b)

$-H^+$

(c)

$-$phosphate

(d)

(e)

H^+

(f)

H_2O
$-H^+$

(g)

H^+

transamination

Threonine
+
Pyridoxal - enzyme imine

(h)

$$R{-}S{-}CH_2CH_2CH(\overset{+}{N}H_3)CO_2^- + \text{N-methylmaleimide} \xrightarrow[\text{pyridoxal}]{\text{enzyme}} R{-}S{-}CH_2CH(\text{succinimidyl})CH(\overset{+}{N}H_3)CO_2^- \quad (3.20)$$

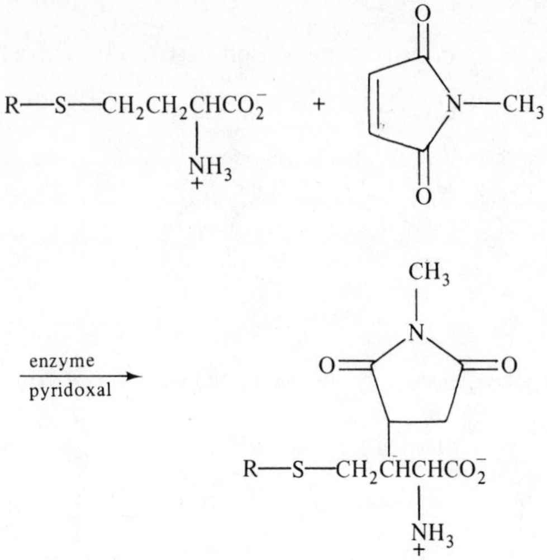

reacts in a Michael reaction (3.20) with N-methylmalonimide, a powerful electrophile, to give the addition product. The deuterium exchange at the gamma carbon predicted from (3.19e → f) is also observed.

Note that the pyridoxal Schiff base stabilizes intermediates with anionic character at the alpha, beta, and gamma carbons. Conjugation with the ring nitrogen stabilizes negative charge at the alpha and gamma positions (3.19a, 3.19e). The imine nitrogen stabilizes negative charge at the beta position, as indicated in (3.19c).

$$RCHCO_2^-\text{–Schiff base} \xrightarrow{-CO_2} R\bar{C}H\text{–Schiff base} \longleftrightarrow RCH{=}\text{quinonoid} \xrightarrow{H^+} RCH_2\text{–Schiff base} \longrightarrow RCH_2NH_2 + \text{Pyridoxal-enzyme imine} \quad (3.21)$$

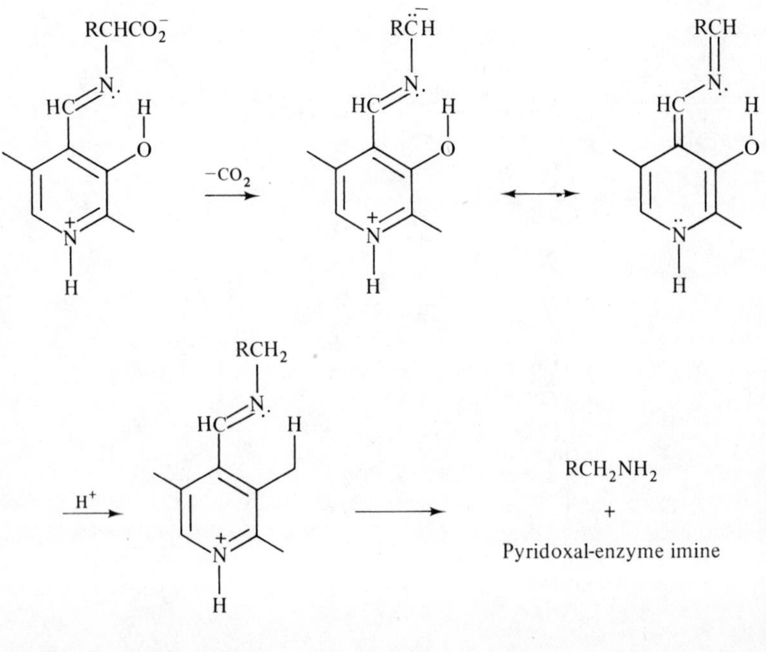

The elimination of carbon dioxide (3.4) and of formaldehyde (3.5) from appropriate amino acids also depends on the ability of a Schiff base with pyridoxal to stabilize a negative charge on the alpha carbon of the substrate. In (3.21), it is the loss of CO_2 rather than a proton that yields the carbanion.

In the fifth reaction, a hydroxymethyl group is reversibly transferred to glycine from tetrahydrofolic acid (formaldehyde is not formed in the free state). The mechanism, given in (3.22), again involves a resonance stabilized carbanion as an intermediate.

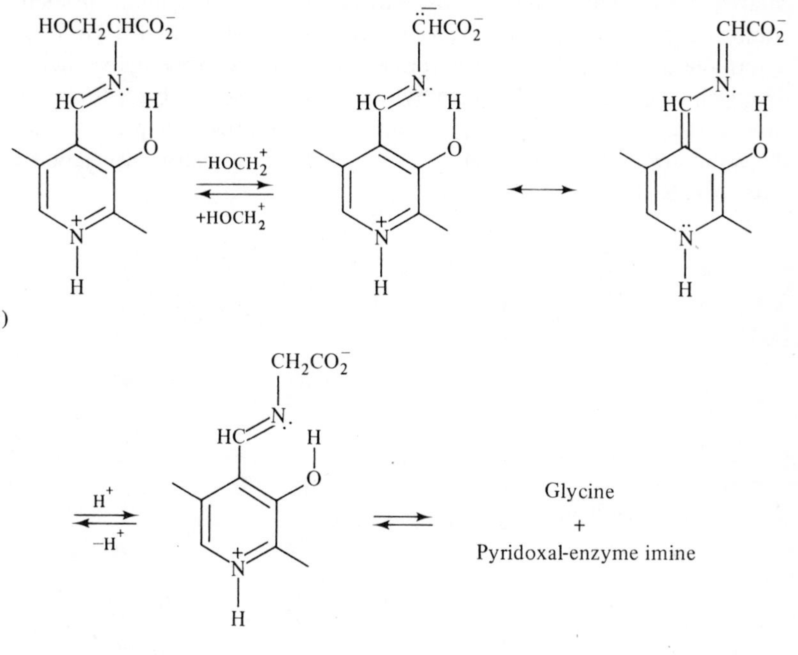

(3.22)

Glycine
+
Pyridoxal-enzyme imine

3.4 STEREOCHEMICAL CONTROL OF PRODUCT FORMATION

At this point, we can note that from an appropriate amino acid bound through nitrogen, enzymes exist that remove the R group, CO_2, or, most usually, a proton. It is important to understand what structural features determine which bond will be broken. Cleavage occurs through the lowest energy transition state. Since each of three bonds may be broken depending on the enzyme, it is not the substrate structure that is rate determining. The conformation of the coenzyme-bound substrate on the enzyme is thought to be the dominant influence. As the sigma bond is broken, the π delocalization is increased. For maximum overlap in the transition state, the bond to the leaving group should be parallel to the p orbitals of the planar Schiff base. Favorable conformations for elimination from

(3.23)

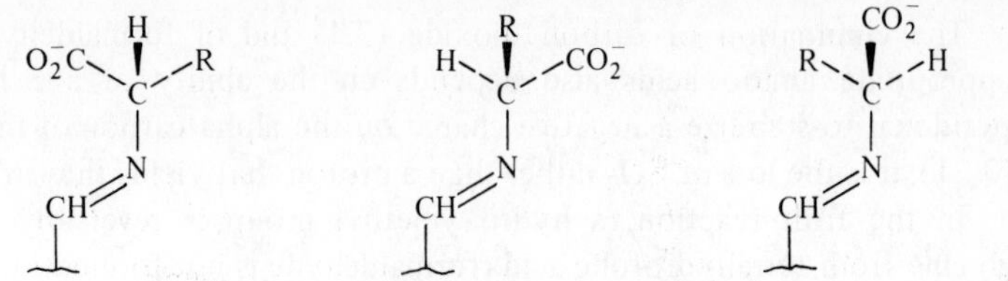

L-amino acids are given in (3.23). (Conformations in which the alpha carbon is rotated 180° would be equally satisfactory for maximum overlap.) The combined ionic, polar, and hydrophobic interactions with the enzyme would determine which conformer was formed. There is experimental support for this description. L-Serine hydroxymethyl transferase also catalyzes the transamination of the D-serine. If the orientation of the carboxyl group is determined by the structure of the binding site, the stereochemical course of reaction follows directly (3.24).

L-Serine

(3.24)

D-Serine

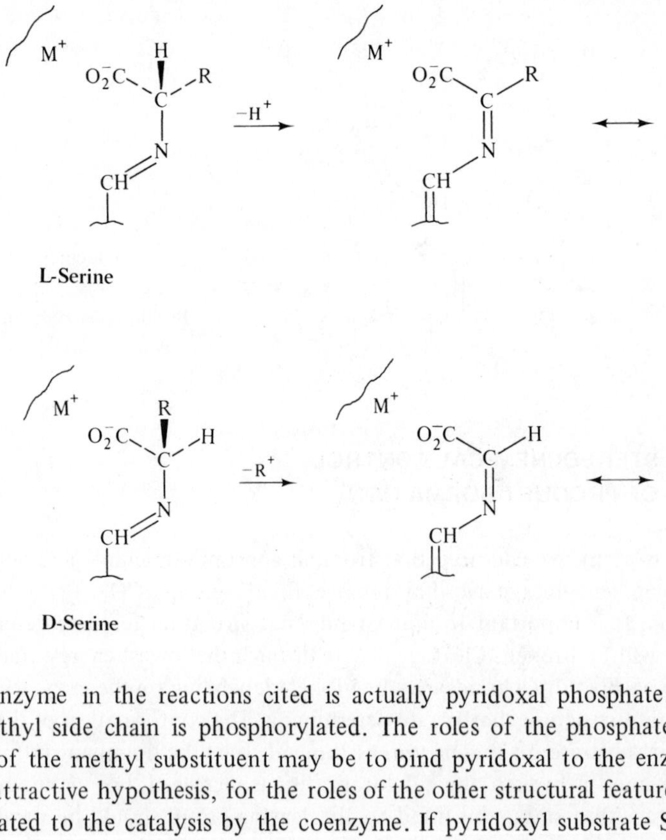

The coenzyme in the reactions cited is actually pyridoxal phosphate. The hydroxymethyl side chain is phosphorylated. The roles of the phosphate side chain and of the methyl substituent may be to bind pyridoxal to the enzyme. This is an attractive hypothesis, for the roles of the other structural features are directly related to the catalysis by the coenzyme. If pyridoxyl substrate Schiff bases were bound to the enzyme as pictured in (3.25), there would be an axis

M++

O− + O− P HO O N: CH H O H₂C O CH₃ +N H

Hydrophobic

(3.25)

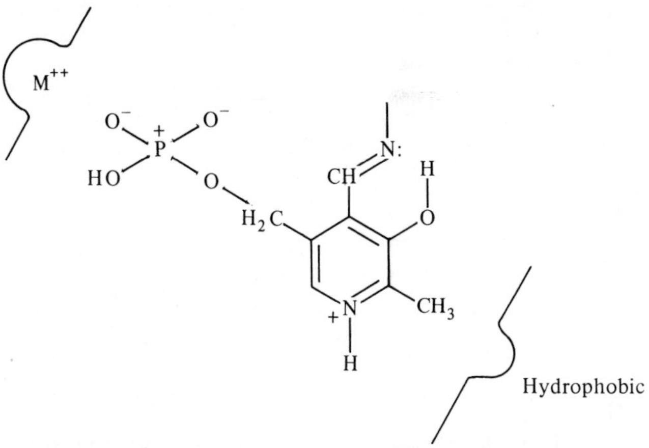

about which pyridoxal could pivot between enzyme-imine and substrate-imine covalent structures. If everything in nature exists for a reason, this may indeed be the case. Certainly past speculation about details of the role of pyridoxal (and other coenzymes) in mechanisms has been a fertile approach to the field of reaction mechanisms.

REFERENCES

Transamination

J. A. Anderson and P. S. Song, *Arch. Biochem. Biophys.*, **122**, 224 (1967).
J. E. Ayling, H. C. Dunathan, and E. E. Snell, *Biochem.*, **7**, 4537(1968).
E. H. Cordes and W. P. Jencks, *Biochem.*, **1**, 773 (1962).
H. C. Dunathan, *Adv. Enz.*, **35**, 79 (1971).
H. C. Dunathan, *Proc. Nat. Acad. Sci.*, **55**, 712 (1966).
P. Fasella, *Ann. Rev. Biochem.*, **36**, 185 (1967).
R. M. Herbst, *Adv. Enz.*, **4**, 75 (1944).
R. C. Hughes, W. T. Jenkins, and E. H. Fischer, *Proc. Nat. Acad. Sci.*, **48**, 1615 (1962).
V. I. Ivanov and M. Y. Karpeisky, *Adv. Enz.*, **32**, 21 (1969).
J. F. Maley and T. C. Bruice, *J. Am. Chem. Soc.*, **90**, 2843 (1968).
J. W. Thanassi, A. R. Butler, and T. C. Bruice, *Biochem.*, **4**, 1463 (1965).

Pyridoxal Eliminations

M. Flavin, *J. Biol. Chem.*, **240**, PC2759 (1965).
M. Flavin and C. Slaughter, *Biochem.*, **3**, 885 (1964).
S. Guggenheim and M. Flavin, *Biochem. Biophys. Acta*, **151**, 664 (1968).
S. Guggenheim and M. Flavin, *J. Biol. Chem.*, **244**, 6217(1969).
M. Krongelb, T. A. Smith, and R. H. Abeles, *Biochem. Biophys. Acta*, **167**, 473 (1968).

4 ACETYL COENZYME A AND LIPOIC ACID; SULFUR IN COENZYME CHEMISTRY

4.1 INTRODUCTION

In this chapter, we consider a molecule that plays a central role in the conversion of food to energy, a molecule that is a key intermediate in the biosynthesis of new larger molecules. Carbohydrates, fats, and proteins are catabolyzed to give acetyl coenzyme A (Acetyl CoA, Acetyl-S-CoA). Acetyl CoA is a thiol ester (4.1), and as a thiol ester it undergoes two types of reactions. Reactions may occur at the methyl group. In these reactions a proton is lost, and the resulting carbanion condenses with an electrophile. Reactions may also occur at the acyl carbon. In these, the CoA molecule is lost, and the acetyl group is transferred to a new molecule. The first type of reaction is responsible for one method of forming new carbon–carbon bonds; the second type is responsible for acetylation reactions. Thus acetyl coenzyme A is both the biological reagent used for Claisen-type condensations and a biological acetylating agent corresponding to acetyl chloride or acetic anhydride. There are many other acyl derivatives of

(4.1)

$$CH_3C(=O)-S-R$$

coenzyme A that occur in the cell (such as succinyl CoA, malonyl CoA, butyryl CoA, etc.). Their chemistry is comparable to that of acetyl coenzyme A and as such will not be discussed in detail.

Acetyl CoA is a transport molecule carrying activated two-carbon units between enzyme complexes used in degradation and enzyme complexes used in synthesis. It carries acyl groups to membranes where it donates them to membrane transfer molecules. On the other side of the membrane carrier, coenzyme A receives the acyl groups and distributes them to enzymes. In cells as in the chemical industry, plants that produce raw materials are more often in different locations than those that produce consumer products.

An analogy might be made between acetyl CoA in a cell and propylene in the chemical industry. Propylene is produced in the cracking of oil. It is used in the production of gasoline, detergents, acetone and phenol via cumene, and of polypropylene using cationic polymerization. It is also burned for its caloric content. The particular use depends upon demand for products, alternate sources of raw materials, availability, and cost of transportation. Acetyl CoA is still more versatile. The particular use of this compound depends upon the needs of the organism. It may be used for fuel or for synthesis. Without giving details, the central position of acetyl coenzyme A in metabolism can be summarized in (4.2).

(4.2)

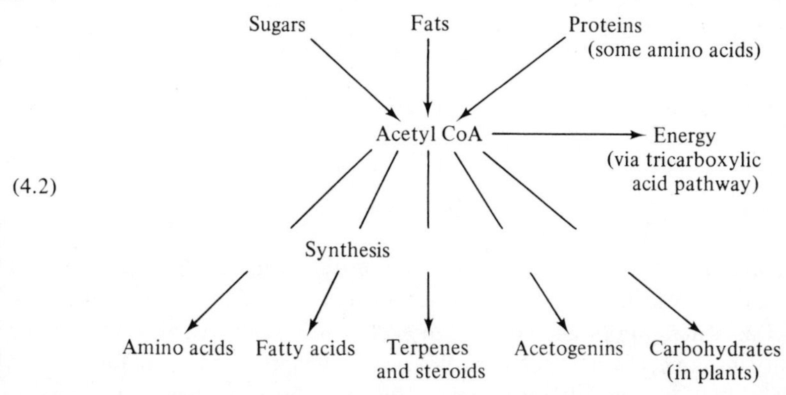

In this chapter, we will discuss the structure of acetyl coenzyme A, its free energy of hydrolysis, and the influence of *d* orbitals of sulfur on the high reactivity at the alpha carbon of the carbonyl. The mechanism for formation of acetyl CoA from pyruvate will be given. Cell strategies for degradation of fatty acids and for their synthesis will be compared and contrasted. Finally, some structure-reactivity relationships of lipoic acid, a coenzyme involved in the formation of acetyl CoA or malonyl CoA, will be presented briefly.

4.2 STRUCTURE AND REACTIVITY OF ACETYL COENZYME A

The structure of coenzyme A is given in (4.3). Coenzyme A has a polarizable strongly nucleophilic sulfhydryl group at the end of a long, relatively polar, water soluble chain. Acetyl CoA has a moderate free energy of hydrolysis. The standard free energy of hydrolysis of acetyl CoA is -7.5 kcal/mole at pH 7; that of ethyl acetate is -4.7 kcal/mole. The transfer of an acetyl group from coenzyme A to an alcohol is strongly favored at equilibrium (i.e., acetyl CoA is a good acetylating agent). Just as the iodide ion is both a better nucleophile and better leaving group than the chloride ion, a mercaptan is a better nucleophile and a better leaving group than an alcohol.

(4.3)

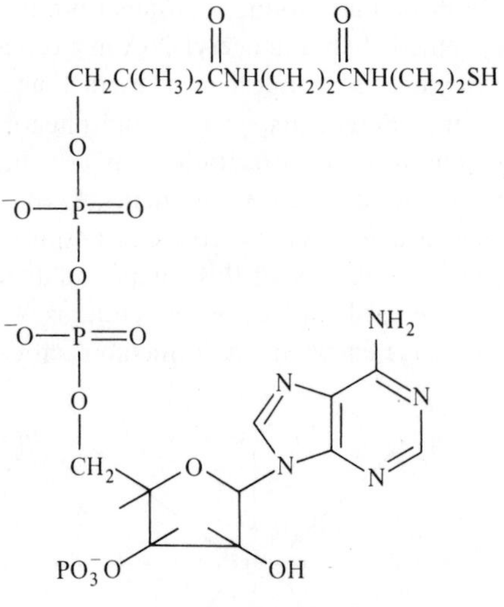

$$CH_2C(CH_3)_2\overset{\overset{\displaystyle O}{\|}}{C}NH(CH_2)_2\overset{\overset{\displaystyle O}{\|}}{C}NH(CH_2)_2SH$$

$$|$$

$$O$$

$$|$$

$$^-O{-}P{=}O$$

$$|$$

$$O$$

$$|$$

$$^-O{-}P{=}O$$

$$|$$

$$O$$

$$|$$

$$CH_2 \text{ (ribose, with } PO_3^-O \text{ and } OH\text{; adenine with } NH_2, N, N, N, N\text{)}$$

Coenzyme A

The electronic structure of thiol esters differs significantly from that of compounds in which the acyl group is bonded to a second-row element. A detailed consideration of the bonding in acetyl CoA is important to identify its reactivity in condensation reactions. Physical evidence for the structure of this ester is obtained from infrared spectra. The stretching frequency of a bond is proportional to the square root of the bond force constant. For a given type of bond, the higher the frequency, the stronger the bond. The carbonyl stretching frequencies for a number of acyl compounds together with the C–H stretching frequency of hydrogen bonded phenylacetylene (4.4) are given in Table 4.1.

(4.4)

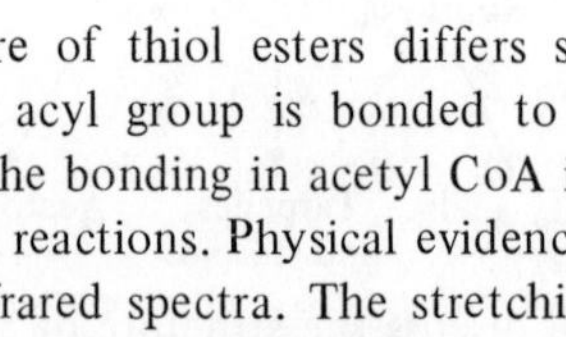

$$\begin{matrix} R \\ \backslash \\ C{=}\ddot{O}{:}{-}{-}H{-}C{\equiv}CC_6H_5 \\ / \\ Y \end{matrix}$$

TABLE 4.1

Compound	$\bar{\nu}_{c=O}$ (cm^{-1})	$\bar{\nu}_{c-H}$ (cm^{-1})
$CH_3C(=O)Cl$	1807	3297
$CH_3C(=O)OCH_3$	1748	3267
$CH_3C(=O)CH_3$	1718	3257
$CH_3C(=O)SCH_3$	1698	3273

Taken from A. W. Baker and G. H. Harris, *J. Am. Chem. Soc.*, **82**, 1923 (1960).

The carbonyl stretching frequency reflects the balance between inductive effects and resonance effects. In esters and acid chlorides, the carbonyl oxygen is bound more firmly than in acetone. The electronegative substituents withdraw electron density inductively (4.5), making the carbonyl carbon more positive.

(4.5) $RC(=O) \leftarrow Y$

The π electrons are bound closer, shortening the carbon–oxygen bond. Even though oxygen is more electronegative than chlorine, the carbonyl stretching frequency of esters is lower than that of acid chlorides. This order reflects the greater importance of the dipolar canonical structure in the ester hybrid compared to that in acetyl chloride (4.6). The low carbonyl stretching frequency for the thiol ester must reflect the lower electronegativity of sulfur relative to oxygen or chlorine and to resonance structures having carbon–oxygen single bond character. Sulfur is more electronegative than carbon, so resonance must be important since acetone has a stronger double bond than the thiol ester.

The bonding in thiol esters is more complex than that inferred from the carbonyl frequencies. If the π electrons are tightly held, the lone pair electrons on oxygen should also be more tightly held. Oxygen should be less basic and

$$CH_3C(=O)OCH_3 \longleftrightarrow CH_3C(O^-)=\overset{+}{O}CH_3$$

(a) (b)

(4.6)

$$CH_3C(=O)Cl \longleftrightarrow CH_3C(O^-)=\overset{+}{Cl}$$

(c) (d)

form weaker hydrogen bonds. For the first three compounds in Table 4.1, this is the case. The stronger the carbonyl bond, the stronger the C–H bond of phenylacetylene and the weaker the hydrogen bond to oxygen. However, methyl thioactate appears to be out of order. The oxygen is more positive (weaker hydrogen bond) than would be predicted from the carbonyl stretching frequency. We are faced with the dilemma of a low bond order in the carbonyl group of a thiol ester, yet a less basic carbonyl oxygen compared to an oxygen ester.

The explanation for the unique features of the thiol ester is found in the ability of sulfur to expand its valence shell by using *d* orbitals. (Sulfur may have a valence of six, as in SF_6.) The thiol ester is a resonance hybrid (4.7) of canonical forms that place both positive and negative charges on oxygen.

$$CH_3C(=O)SCH_3 \longleftrightarrow CH_3C(O^-)=\overset{+}{S}CH_3 \longleftrightarrow CH_3C(O^+)=\overset{-}{S}CH_3 \quad (4.7)$$

(a) (b) (c)

Structures (4.7b) and (4.7c) contribute to lengthening and weakening the carbonyl bond of the thiol ester. Because these structures are opposed in charge, the lowering of the carbonyl frequency is not accompanied by increased basicity of oxygen. More needs to be said about both dipolar structures in (4.7). Structure (4.7b) represents the overlap of $3p$ orbitals on sulfur with $2p$ orbitals on carbon and oxygen. This overlap is small: (4.7b) contributes much less to the hybrid of the thiol ester than the corresponding structure does to an ester. Second-row elements are good double bond formers, but third-row elements are notoriously poor. (4.7c) has no precedence in compounds formed between second row elements. A 3*d* orbital on sulfur has the proper symmetry to overlap with a 2*p* orbital (4.8). This overlap increases the electron density on sulfur and reduces the density on the carbonyl group. A molecular orbital description of a thiol ester avoids the misleading electron deficient oxygen in (4.7c).

(4.8)

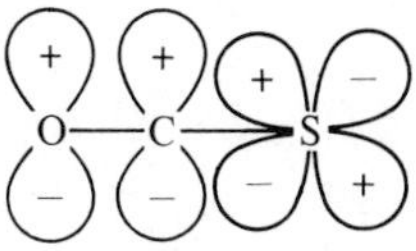

O—C—S (orbital lobes: + / − on O; + / − on C; + / − and − / + on S)

Expansion of sulfur's valence shell by *d*-orbital participation is also helpful in explaining reactions of thiol esters at the alpha position. Ethyl thiol acetate (4.9a) undergoes exchange with D_2O at more than ten times the rate of ethyl acetate (4.9b). The sulfur analogue of ethyl acetoacetate is a stronger acid by more than two pK_a units (4.10). Such differences in acidity are exceedingly important for biochemical reactions, for strong acids and bases are not available in cells. With its increased acidity, acetyl CoA is a potential nucleophilic carbanion for the formation of new carbon–carbon bonds.

(4.9)

(a)	(b)
$CH_3C(=O)SC_2H_5$	$CH_3C(=O)OC_2H_5$
Relative rate = 13	Relative rate = 1

(4.10)

$CH_3\overset{O}{\overset{\|\|}{C}}CH_2\overset{O}{\overset{\|\|}{C}}SC_2H_5$	$CH_3\overset{O}{\overset{\|\|}{C}}CH_2\overset{O}{\overset{\|\|}{C}}OC_2H_5$
$pK_a = 8.50$	$pK_a = 10.70$

The extra acidity of the methyl group of acetyl CoA may be explained in two different ways that essentially describe the same phenomena. The more positive carbonyl oxygen of a thiol ester is a better electron sink for the enol (that is, the sulfur adds another electron sink for the enol). The resonance structures of this nucleophile are given in (4.11). To the extent that (4.7b) and (4.7c) make smaller contributions to thiol esters than (4.6b) does to oxyesters, (4.11d) makes a greater contribution to the thiol carbanion than (4.12c) does to the corresponding oxygen analogue. This gain in delocalization accompanying ionization of acetyl CoA facilitates proton removal and stabilizes the carbanion.

(4.11)

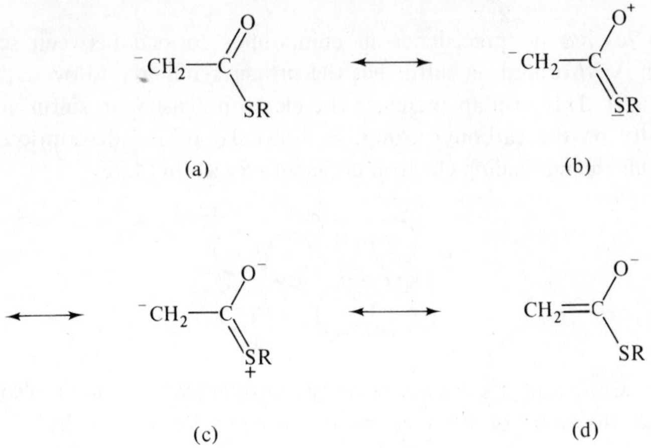

$$^{-}CH_2{-}C({=}O){-}SR \;\longleftrightarrow\; ^{-}CH_2{-}C(O^{+}){=}\underline{S}R \;\longleftrightarrow\; ^{-}CH_2{-}C(O^{-}){=}S^{+}R \;\longleftrightarrow\; CH_2{=}C(O^{-}){-}SR$$

(a) (b) (c) (d)

(4.12)

$$^{-}CH_2C({=}O){-}OR \;\longleftrightarrow\; ^{-}CH_2C(O^{-}){=}O^{+}R \;\longleftrightarrow\; CH_2{=}C(O^{-}){-}OR$$

(a) (b) (c)

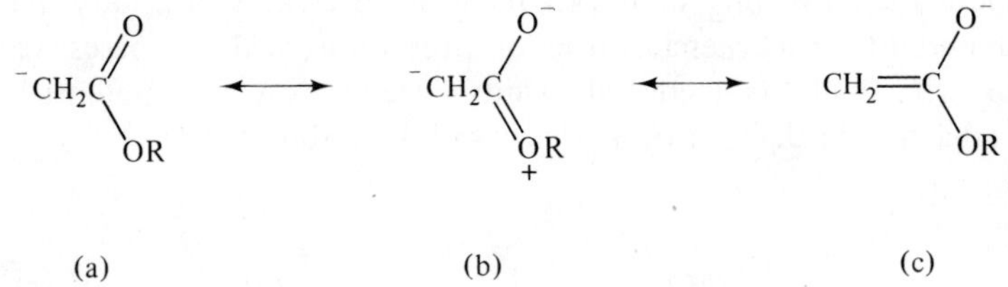

4.3 THE FORMATION OF ACETYL CoA FROM PYRUVATE

Acetyl CoA is the source of two-carbon fragments for the Krebs tricarboxylic acid cycle. This is the final common oxidative pathway in aerobic cells. The net reaction of the two-carbon fragments is given in (4.13). Carbon dioxide is the oxidation product; NADH and reduced flavins are the reduction products. The latter high-energy compounds may be oxidized in enzymes that also produce high-energy ATP.

(4.13) $$CH_3COOH + 2H_2O \rightarrow 2CO_2 + 8H$$

Acetyl CoA fuel can be derived from coenzyme A esters of β-keto acids by reverse condensation reactions (Section 4.4). Some amino acids and all fatty acids follow this pathway. Other amino acids and carbohydrates enter the Krebs cycle via pyruvate. Sugars are first converted to pyruvate. Pyruvate is then converted to acetyl CoA, carbon dioxide, and reduced NADH (4.14) in reactions involving a variety of coenzymes. The mechanisms in this conversion are the subject of this section.

(4.14)

$$NAD^+ + RSH + CH_3\overset{O}{\overset{\|}{C}}CO_2^- \rightarrow NADH + CH_3\overset{O}{\overset{\|}{C}}SR + CO_2$$

In the first step of the conversion, thiamine pyrophosphate adds to pyruvate (4.15). The positive nitrogen acts as an electron sink to stabilize the carbanion (4.16b) resulting from the decarboxylation of the adduct. The enamine (4.17a) adds to lipoic acid (4.17). Thiamine adds as a nucleophile and departs as a nucleophile. (The structure-function relationships that contribute to the suitability of thiamine for this role are discussed in Chapter 5.)

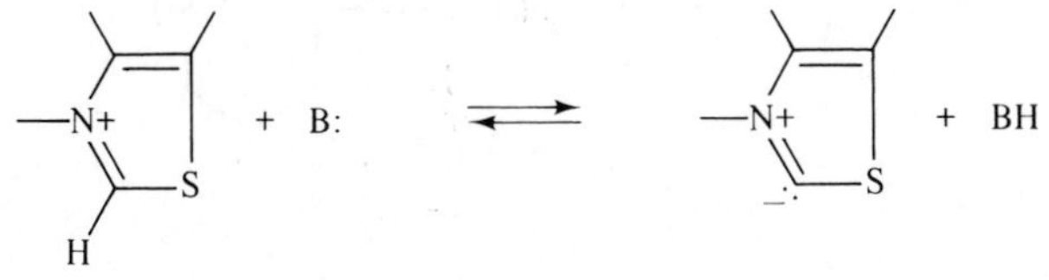

(4.15)

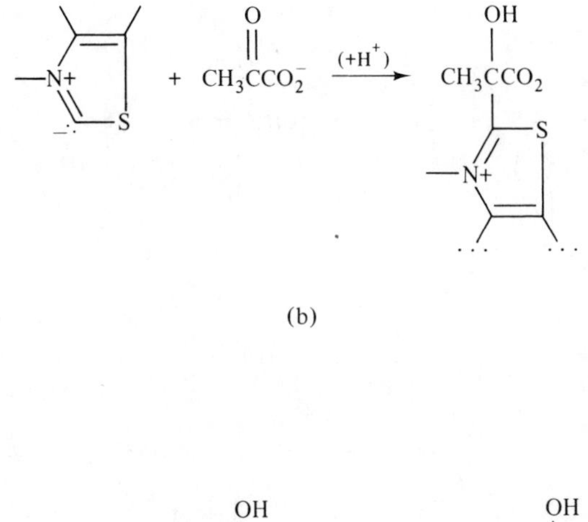

—N+ ... S, H + B: ⇌ —N+ ... S + BH

(a)

—N+ ... S + $CH_3\overset{O}{\overset{\|}{C}}CO_2^-$ $\xrightarrow{(+H^+)}$ $CH_3\overset{OH}{\overset{|}{C}}CO_2$ —N+ ... S

(b)

(4.16)

$CH_3\overset{OH}{\overset{|}{C}}CO_2^-$ (—N+, S) $\xrightarrow{-CO_2}$ $CH_3\overset{OH}{\overset{|}{C}}:^-$ (—N+, S) ⟷ $CH_3\overset{OH}{\overset{|}{C}}$ (—N:, S) $\xrightarrow{H^+}$ $CH_3\overset{OH}{\overset{|}{C}}H$ (—N+, S)

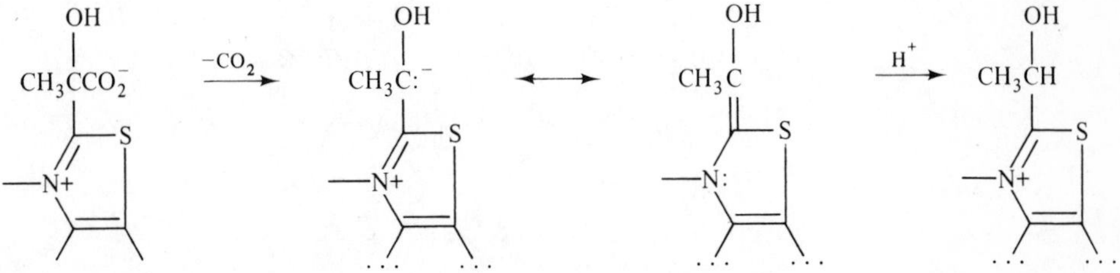

(a) (b) (c)

(4.17)

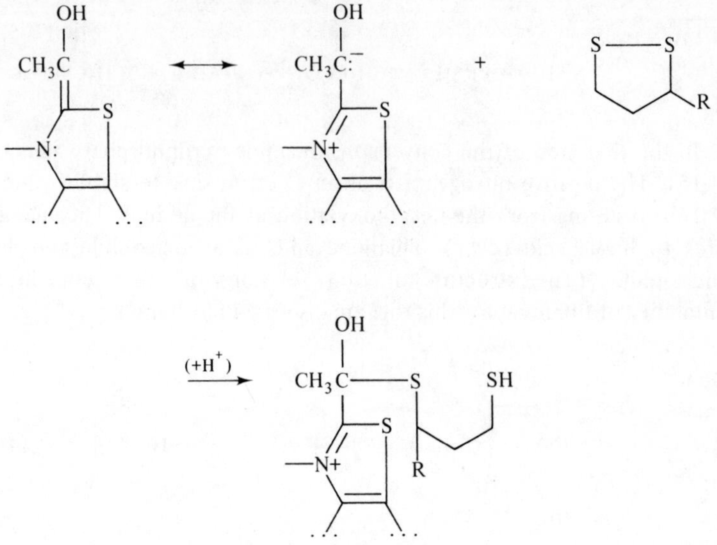

(a)

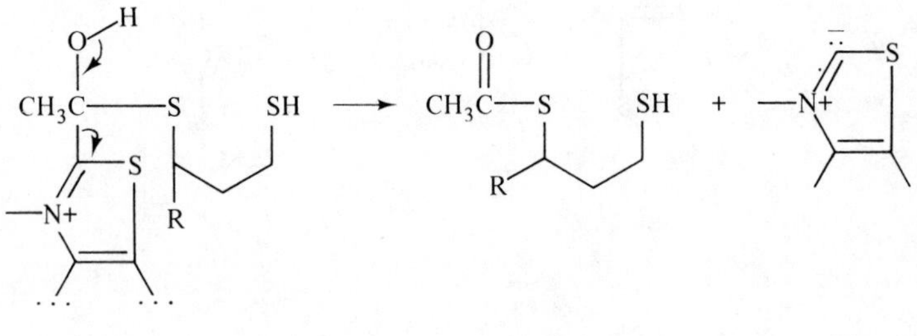

(b)

(4.18)

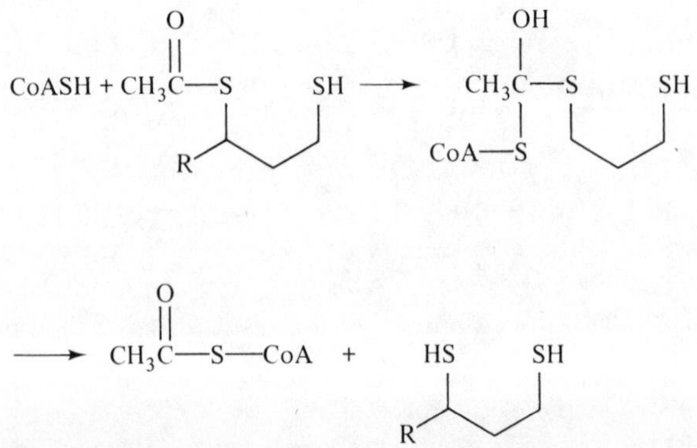

$$CoASH + CH_3C(=O)\text{—}S\text{—}(R)CH\text{—}CH_2\text{—}CH_2\text{—}SH \longrightarrow CH_3C(OH)(S\text{—}CoA)\text{—}S\text{—}\ldots\text{—}SH$$

$$\longrightarrow CH_3C(=O)\text{—}S\text{—}CoA + HS\text{—}(R)CH\text{—}CH_2\text{—}CH_2\text{—}SH$$

The nucleophilic displacement on sulfur (4.17a) is an oxidation of carbon and a reduction of sulfur. This contrasts to nucleophilic displacements on carbon that never involve changes in oxidation states. Lipoic acid is the third coenzyme introduced in this chapter that contains sulfur. It is discussed further in Section 4.5. In a final step, acetyl CoA displaces reduced lipoic acid (4.18). The lipoic acid is then oxidized by NAD^+.

In a series of acid-base reactions and nucleophilic displacement or substitution reactions, pyruvate has been oxidized and converted to acetyl CoA to be used for further oxidation or for synthesis. Formally, the reaction from (4.16a) to (4.16c) results in oxidation of the carboxyl carbon and reduction of the carbonyl carbon of pyruvate. Coenzymes have been used as nucleophiles and leaving groups, and have served to modify the acidity to promote the decarboxylation and nucleophilicity of the substrate.

Acetyl CoA then enters the Krebs cycle by condensation with oxaloacetic acid to give citric acid (4.19).

$$\text{CoA—S}\overset{\overset{\displaystyle O}{\|}}{C}CH_3 \quad + \quad O{=}\underset{\underset{\displaystyle CO_2^-}{|}}{C}CH_2CO_2^-$$

$$(4.19) \qquad \longrightarrow \quad \text{CoA—S}\overset{\overset{\displaystyle O}{\|}}{C}CH_2\underset{\underset{\displaystyle CO_2^-}{|}}{\overset{\overset{\displaystyle OH}{|}}{C}}CH_2CO_2^- \quad \longrightarrow$$

$$^-O\overset{\overset{\displaystyle O}{\|}}{C}CH_2\underset{\underset{\displaystyle CO_2^-}{|}}{\overset{\overset{\displaystyle OH}{|}}{C}}CH_2\overset{\overset{\displaystyle O}{\|}}{C}O^- \quad + \quad \text{CoASH}$$

4.4 CELL STRATEGIES FOR SYNTHESIS AND DEGRADATION

Synthetic chemists and industrial chemists are frequently concerned with improving the yield of reactions. In a ten-step synthesis in which each reaction proceeds with 90% yield, the overall yield is only 35%. With substantially lower yields, a long synthesis may be impractical or even impossible. Low yields in reactions also mean greater amounts of impurities. Separations become less efficient and more costly in time and effort. A cell faces these same problems.

Three economies need to be maximized. First, intermediates should not accumulate. In a cell, unneeded compounds serve only to dilute needed compounds and to increase the entropy of the system. A living system uses energy to create order. This task is sufficiently challenging without the added hazard of poor planning and technique. A second economy concerns genetic material. Each different enzyme is coded by about four times its molecular weight in DNA. If the same enzymes could be used for degradation and for synthesis, the cell would require less DNA. In the synthesis of glucose, eight of the eleven enzymes used in its breakdown to lactate (glycolysis) are used together with some completely different enzymes. These economies are not always compatible. Enzymes that favor bond breakage at equilibrium require a large excess of starting materials for synthesis. The three glycolytic enzymes bypassed in synthesis belong to this category. The third area of economy might be called plant design. The enzymes for fatty acid synthesis are associated. The growing acid is bound to a long flexible chain and passed from one production step to another with minimum transportation costs. (The association of protein molecules in enzyme clusters is referred to as the quaternary structure of proteins.)

The reactions for the degradation of fatty acids follow with some brief comments. Important differences between fatty acid synthesis and degradation are noted. Fatty acids are converted to acetyl CoA by the scheme given in (4.20) through (4.24). The reactions given in (4.20) introduce the desired thiol ester group that appears in acetyl CoA. Because the hydrolysis of acetyl CoA has a large negative ΔG, the reaction is coupled to the hydrolysis of two bonds of ATP to force the equilibrium to the right. The ATP also acts to provide a mechanism for the reaction. The carboxylate is converted to an acid anhydride (4.25); the diphosphate serves as a leaving group in a nucleophilic substitution reaction. This activation of a carboxylate is not required for synthesis reactions: acetyl CoA is the initial building block.

$$RCH_2CH_2CO_2^- + ATP + CoASH \longrightarrow$$

$$RCH_2CH_2C(=O)SCoA + AMP + PP_i \qquad (4.20)$$

(a)

$$H_2O + PP_i \rightarrow 2P_i$$

(b)

$$RCH_2CH_2C(=O)SCoA + FAD \longrightarrow RCH{=}CHC(=O)SCoA + FADH_2 \tag{4.21}$$

$$RCH{=}CHC(=O)SCoA + H_2O \longrightarrow RCH(OH)CH_2C(=O)SCoA \tag{4.22}$$

$$RCH(OH)CH_2C(=O)SCoA + NAD \longrightarrow RC(=O)CH_2C(=O)SCoA + NADH + H^+ \tag{4.23}$$

$$RC(=O)CH_2C(=O)SCoA + CoASH \longrightarrow RC(=O)SCoA + CH_3C(=O)SCoA \tag{4.24}$$

$$\text{RCH}_2\text{CH}_2\overset{\overset{\displaystyle O}{\|}}{C}\text{—O}\overset{\overset{\displaystyle O}{\|}}{\underset{\underset{\displaystyle O^-}{|}}{P}}\text{O}\overset{\overset{\displaystyle O}{\|}}{\underset{\underset{\displaystyle O^-}{|}}{P}}\text{OH} \tag{4.25}$$

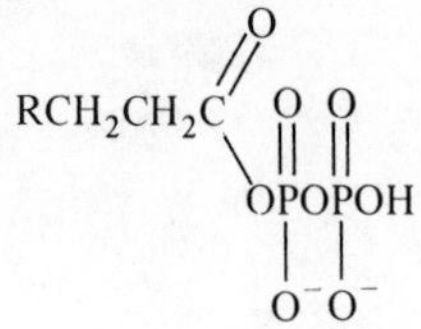

Reactions (4.21) through (4.22) introduce a β-keto function. We may speculate on (4.21). Carbanions, because of their high electron density, are easily oxidized by losing one electron to free radicals. Flavins, discussed in Chapter 5, are good one-electron oxidizing agents. The thiol ester would increase the acidity of the alpha proton allowing a carbanion to form which would subsequently be oxidized by the flavin. The thiol ester may also stabilize the free radical intermediate (4.26). Because of the dipolar resonance of the α, β-unsaturated thiol ester (4.27), the water adds to the beta position. The resonance forms described by (4.27) are more important in a thiol ester than in an oxygen ester,

$$\text{R—}\dot{\text{C}}\text{H—C}(=\ddot{\text{O}}\text{:})\text{SCoA} \longleftrightarrow \text{RCH}{=}\text{C}(\text{—}\ddot{\text{O}}\text{:})\text{SCoA} \tag{4.26}$$

$$\text{RCH}{=}\text{CH—C}(=\text{O})\text{SCoA} \longleftrightarrow \text{R}\overset{+}{\text{C}}\text{H—CH}{=}\text{C}(\text{O}^-)\text{SCoA} \tag{4.27}$$

$$\text{NADPH} + \text{RCH}{=}\text{CHC}(=\text{O})\text{SCoA} \longrightarrow$$

$$\text{NADP}^+ + \text{RCH}_2\ddot{\overset{-}{\text{C}}}\text{HC}(=\text{O})\text{SCoA} \longleftrightarrow \text{RCH}_2\text{CH}{=}\text{C}(\text{O}^-)\text{SCoA} \tag{4.28}$$

$$\xrightarrow{\text{H}^+} \text{RCH}_2\text{CH}_2\text{C}(=\text{O})\text{SCoA}$$

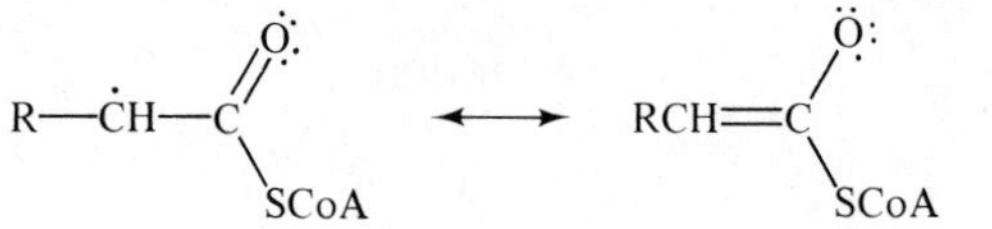

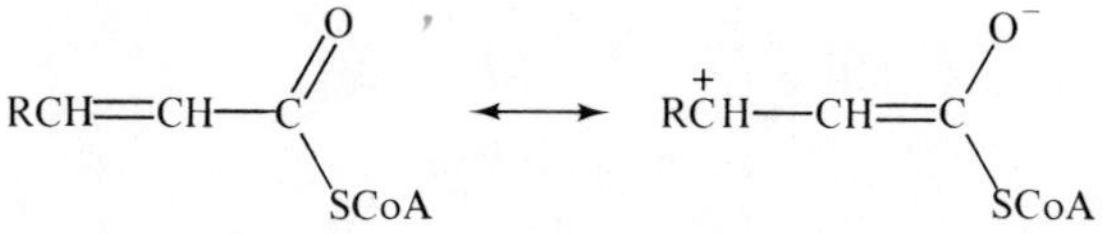

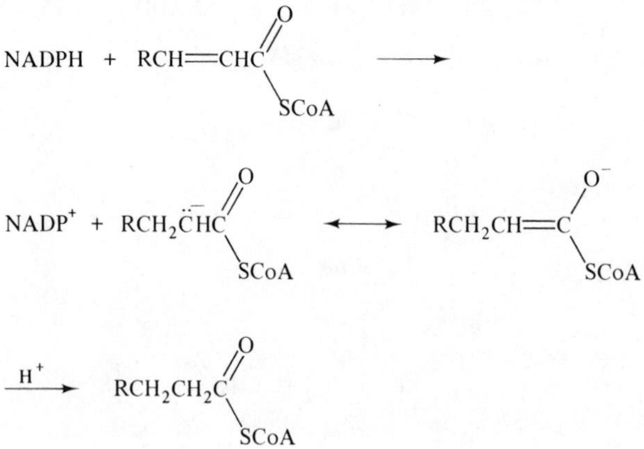

because the carbonyl oxygen in a thiol ester is more positive and therefore a better electron sink. NAD^+ then removes a hydride ion in forming the β-keto ester. Similar steps occur in the reduction of a β-keto ester intermediate in the formation of fatty acids. In contrast, however, the hydrogenation of the double bond occurs by a polar mechanism (4.28), where again the thiol ester group aids to stabilize the alpha carbanion.

A major contrast in pathways is found in the choice of condensation reactions. The equilibrium for (4.24) lies far to the right. While the reverse reaction is mechanistically satisfactory (4.29), it is not energetically favorable for synthesis. Instead, a pathway involving conversion of acetyl CoA to malonyl CoA, condensation of enzyme-bound thiol esters, and decarboxylation is used [(4.30) through (4.32)]. The addition of CO_2 to acetyl CoA is driven by the hydrolysis of ATP.

$$RC(=O)CH_2C(=O)SCoA + CoAS{-}H \rightleftharpoons RC(O^-)(SCoA)-CH_2C(=O)SCoA \rightleftharpoons$$

(4.29)

$$RC(=O)SCoA + {:}\overline{C}H_2{-}C(=O)SCoA \longleftrightarrow CH_2{=}C(O^-)SCoA \underset{-H^+}{\overset{H^+}{\rightleftharpoons}} CH_3C(=O)SCoA$$

(4.30)

$$CH_3C(=O)SCoA + \text{Biotin-}NCO_2^- + ATP \longrightarrow {}^-OC(=O)CH_2C(=O)SCoA + \text{Biotin (HN–NH)} + ADP + P_i$$

$$\text{Enzyme—S—}\overset{\text{O}}{\overset{\|}{\text{C}}}\text{CH}_3 + \text{Enzyme—S—}\overset{\text{O}}{\overset{\|}{\text{C}}}\text{CH}_2\text{CO}_2^- \longrightarrow \tag{4.31}$$

$$\text{Enzyme—S—}\overset{\text{O}}{\overset{\|}{\text{C}}}\text{CH}_2\overset{\text{O}}{\overset{\|}{\text{C}}}\text{CH}_2\text{CO}_2^- + \text{Enzyme—SH}$$

$$\text{H}^+ + \text{Enzyme—S—}\overset{\text{O}}{\overset{\|}{\text{C}}}\text{CH}_2\overset{\text{O}}{\overset{\|}{\text{C}}}\text{CH}_2\text{CO}_2^- \rightarrow \text{Enzyme—S—}\overset{\text{O}}{\overset{\|}{\text{C}}}\text{CH}_2\overset{\text{O}}{\overset{\|}{\text{C}}}\text{CH}_3 + \text{CO}_2 \tag{4.32}$$

The decarboxylation reaction, by removing the product from (4.31), drives the condensation toward completion. The carboxylation by biotin followed by decarboxylation provides an indirect method to use the hydrolytic energy of ATP to shift the equilibrium toward formation of fatty acids.

4.5 LIPOIC ACID

Lipoic acid (4.33a) and dihydrolipoic acid (4.33b) form a redox couple that can catalyze two electron transfers by heterolytic mechanisms. In the oxidation of pyruvate to acetyl CoA (4.16), lipoic acid serves as an oxidizing agent that conserves part of the energy of oxidation as a thiol ester.

(4.33)

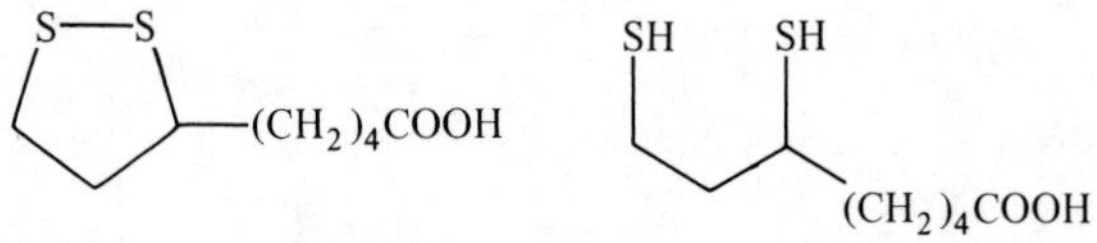

S—S

$-(CH_2)_4COOH$

(a)

SH SH

$(CH_2)_4COOH$

(b)

(4.34)

N⁺

$CONH_2$

B: H—S S—H

→

N

H H $CONH_2$

+

BH + S—S

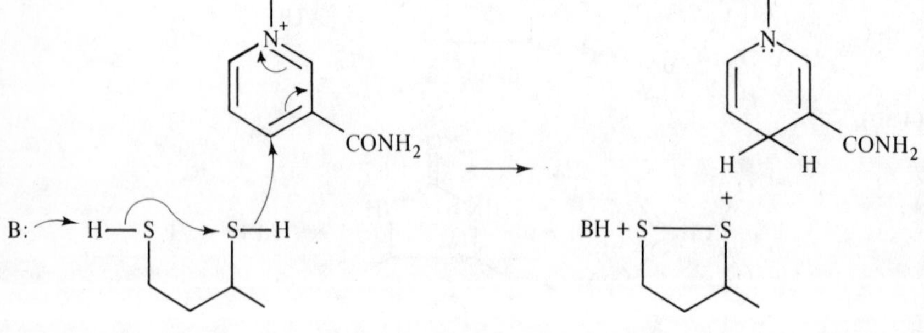

Notice that the mechanism of oxidation by lipoic acid consists of two simple displacement reactions. First there is a displacement by the carbanion which opens the disulfide, and second, coenzyme A displaces the acetyl group from dihydrolipoic acid. After dihydrolipoic acid has been replaced by coenzyme A, the acid is reoxidized by NAD^+. Presumably, this reaction is a hydride transfer reaction (4.34).

An example of a reaction in which lipoic acid acts as a reducing agent occurs in the reduction of sulfate ion to sulfite ion by reduced lipoic acid. The sulfate ion must be first "activated" before attack by the reduced lipoic acid is possible. Two molecules of ATP react with one of the sulfate ion to give 3′-phosphoadenosine-5-phosphosulfate (PAPS). Reduced lipoic acid is now able to attack the sulfate to form a lipothiosulfate and 3′-phospho-AMP (4.35).

(4.35) HS SH (ring) + $O{=}S^{+}(O^-)(O^-)\text{—}O\text{—}P^{+}(O^-)(O^-)\text{—}O\text{—}R$ → HS S–SO_3^- (ring, ^-O, O, O^- on S)

Nucleophilic attack by the other sulfhydryl group will form oxidized lipoic acid and sulfite ion (4.36). Nucleophilic attacks on divalent sulfur of alkyl thiosulfates are well known. For example, the exchange of sulfite ion with alkyl thiosulfates (4.37) has been studied in detail. During these reactions, the lipoic acid is bound to the enzyme as an amide of lysine.

(4.36) HS → S—SO_3^- (ring) ⟶ S—S (ring) + $SO_3^{=}$ + H^+

(4.37) $S^*O_3^{=} \rightarrow \underset{R}{S}\text{—}SO_3^- \rightleftharpoons {}^-O_3S^*\text{—}\underset{R}{S} + SO_3^{=}$

In this section, the structural features that make lipoic acid and dihydrolipoic acid particularly well suited to catalyze these reactions will be presented. Mechanistically, it has already been established that a disulfide and a pair of mercaptan groups are satisfactory. We wish to consider reasons why the redox pair used in the biological reaction reacts faster than alicyclic reagents (4.38).

(4.38) $$2[H] + R\text{—}S\text{—}S\text{—}R \rightarrow 2RSH$$

The oxidation of dihydrolipoic acid is facilitated by having the two thiol groups in the same molecule. If separate mercaptans were to react and form a disulfide, there would be a loss of translational motion: two molecules have joined to become one. The loss of entropy for such a process is unfavorable to both rate and equilibrium. Conformations of a dithiol that bring the terminal sulfurs near to one another are favorable for either three- or four-carbon chains. This ring closure contributes to a lower entropy of oxidation for dihydrolipoic acid relative to single mercaptans.

$$R{-}S^{*}H + R'S{-}SR' \longrightarrow RS^{*}{-}SR' + R'SH \qquad (4.39)$$

That lipoic acid is more reactive than an acyclic disulfide may seem more surprising. This is the case for the exchange reaction (4.39) in Table 4.2. The explanation is not entropic. Indeed the entropy of reaction is unfavorable for the exchange with the cyclic sulfide. For this reaction ΔH is -5.3 kcal/mole, but ΔG is only -1.2 kcal/mole. Lipoic acid is a strained molecule and this strain is released when the ring is opened.

TABLE 4.2

Reactants			k_{ex}	ΔG
$n\text{-}C_4H_9S^{*}H$	+	$n\text{-}C_4H_9SS\text{-}n\text{-}C_4H_9$	0.31	18.1 kcal/mole
$n\text{-}C_4H_9SH$	+	S—S (five-membered ring)	1400	13.6 kcal/mole

If the transition states for the oxidation and reduction of cyclic and acyclic dithiols and disulfides are similar, then the ground-state energies of lipoic and dihydrolipoic acid make them favored for catalysis (4.40). Because of strain, lipoic acid has a higher enthalpy than its acyclic analogue. Because it is a single cyclic molecule, lipoic acid has a lower entropy than two mercaptans.

The strain energy of lipoic acid is about 4.0 kcal/mole. Part of this is due to eclipsing of methylene groups. Part is due to repulsions of electron pairs on adjacent sulfurs (4.41). In H_2S_2, the dihedral angle is 90°. Lipoic acid is not planar: X-ray studies indicate that the dihedral angle, ∠CSSC, is 26½°. We can note that the five-member ring disulfide is a better oxidizing agent than a less strained six-membered ring compound would be.

(4.40)

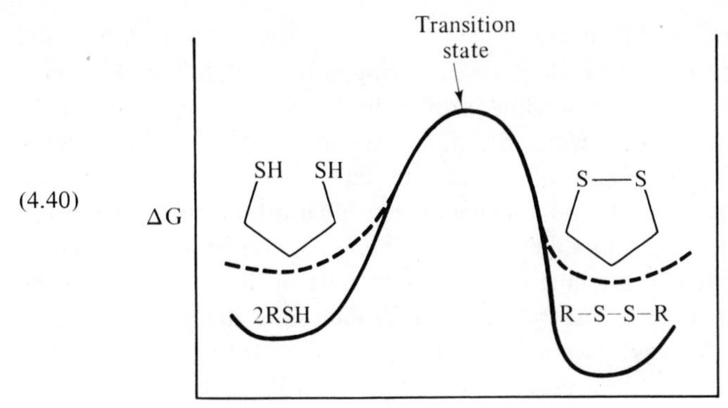

(4.41)

S—S

—$(CH_2)_4COOH$

The electronic repulsion of lipoic acid must be treated in a more sophisticated way to account for spectra. Lipoic acid is yellow; unstrained disulfides are not. If the dihedral angle is 90°, nonbonding electron pairs on sulfurs can be placed in 3*s* orbitals and in orthogonal *p*-orbitals (4.42). Nonbonded electron pairs repel one another more strongly than bonded electron pairs. In the dithiocyclopentane, the lone pair *p* orbitals are parallel. They overlap to form π and π^* orbitals (4.43). These are filled. The fact that the energy of an antibonding π^* orbital is always raised slightly more than the π orbital is stabilized with reference to 3*p* orbitals results in a small net repulsion.

(4.42)

SS —H

H

(4.43)

π π^*

$S_A\,3_p$ — π^* / π — $S_B\,3_p$

We wish to compare the energy needed to photo excite an electron from the highest energy-filled orbital to the lowest energy-empty orbital. The electron is being raised from a π^* antibonding orbital in lipoic acid and from a $3p$ nonbonding orbital in an acyclic disulfide. Assume the high-energy orbital contributes little to the energy of the excited state. Compared to the acyclic molecule, the cyclic disulfide has a higher-energy filled orbital. If an electron is excited from the π^* antibonding orbital, bonding increases (antibonding decreases). This increased bonding in the excited state of lipoic acid lowers the energy of the excited state. The raising of the ground-state energy and lowering of the excited-state energy, diagramed in (4.44), account for the yellow color of lipoic acid.

(4.44)

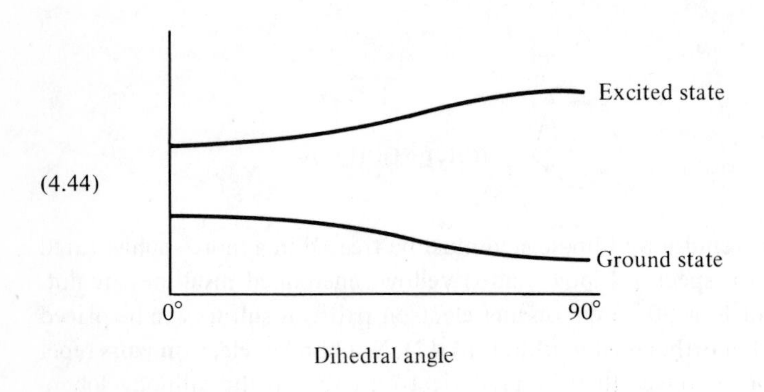

REFERENCES

Coenzyme A

J. Baddiley, *Adv. Enz.*, **16**, 1 (1955).

Lipoic Acid

J. A. Barltrop, P. M. Hayes, and M. Calvin, *J. Am. Chem. Soc.*, **76**, 4348 (1954).
G. Bergson, *Arkiv. Kemi.*, **12**, 233 (1958).
B. Bergson, G. Claeson, and L. Schotte, *Acta Chem. Scand.*, **16**, 1159 (1962).
D. B. Boyd, *J. Am. Chem. Soc.*, **94**, 8799 (1972).
M. Calvin and J. A. Barltrop, *J. Am. Chem. Soc.*, **74**, 6153 (1952).
H. Hilz and M. Kittler, *Biochem. Biophys. Acta*, **30**, 650 (1958).
H. Hilz and M. Kittler, *Biochem. Biophys. Res. Comm.*, **3**, 140 (1960).
L. J. Reed, *Adv. Enz.*, **18**, 319 (1957).
U. Schmidt, P. Grafen, K. Altland, and H. W. Goedde, *Adv. Enz.*, **32**, 423 (1969).

5 COENZYME FUNCTION AND DESIGN: FLAVINS AND THIAMINE

5.1 BIOLOGICAL USES OF FLAVINS

Free-radical reactions, well known in organic chemistry, also occur in biological systems. Typical organic examples of reactions occurring by radical mechanisms include addition polymerizations of vinyl compounds and oxidations of hydrocarbons by chlorine or oxygen (occurring by chain mechanisms). Dissolving metal reductions (Na/EtOH, Na/NH_3) proceed through anion radicals, and certain oxidations in solution using chromate or permanganate proceed through radical intermediates.

Biological analogues do not exist for all these types of radical reactions. Biopolymers are condensation polymers–proteins and nucleic acids have fixed molecular weights in contrast to the varying molecular weights of synthetic polymers. And while chain oxidations do occur in nature, such reactions are unwanted–the oxidation of fats without conserving part of the energy as ATP serves to degrade valuable molecules with a total loss of stored chemical energy.

(5.1)

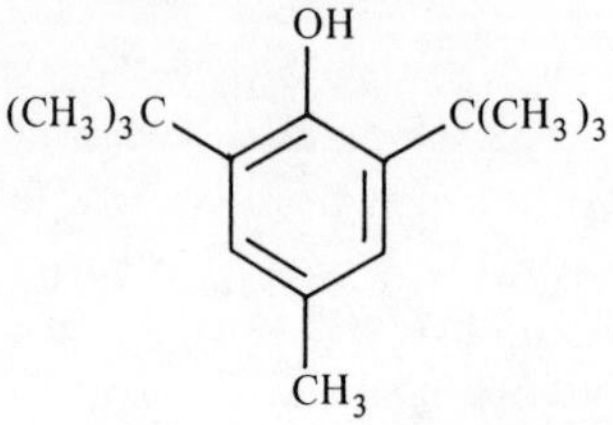

Note too, that whereas radical oxidations are a part of the aging process, radical chain initiators found in smog hasten this destruction. Radicals can act as "traps," and serve as a biological defense mechanism: just as BHT (butylated hydroxytoluene) (5.1) is an additive for edible fats and oils, vitamin E, α-tocopherol (5.2), may perform a similar function in the lungs. The removal of the phenolic hydrogen of either of these compounds gives an unreactive, resonance-stabilized radical terminating the chain reaction (5.3).

(5.2)

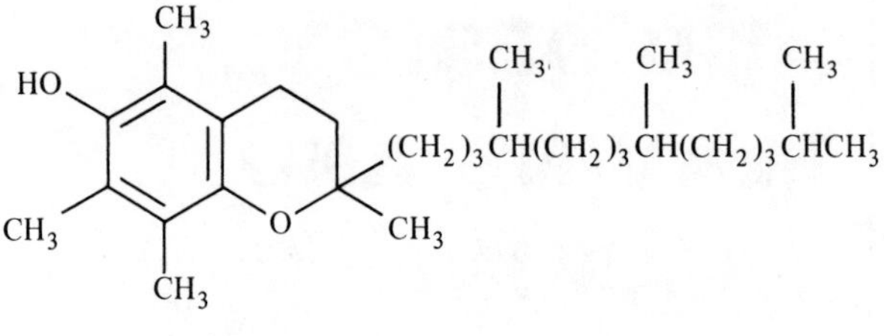

(5.3)

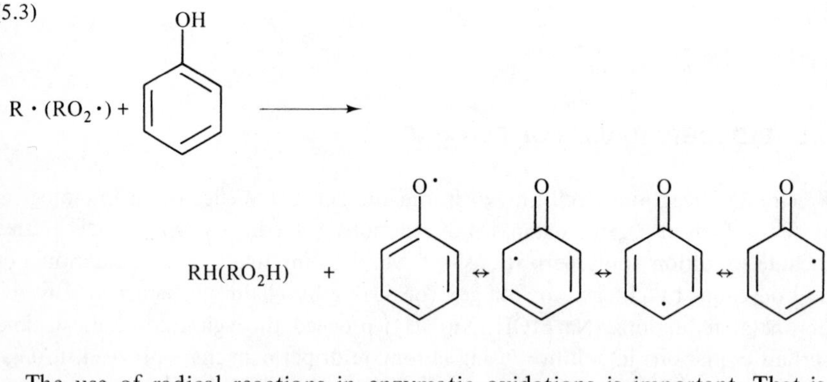

$R\cdot (RO_2\cdot) +$ →

$RH(RO_2H)$ +

The use of radical reactions in enzymatic oxidations is important. That is, since electrons are donated to oxygen through iron porphyrin compounds employing the ferrous-ferric redox couple, and since NADH reacts as a hydride donor (Chapter 1), the body needs a molecule that can serve at the interface of these different types of redox pairs.

Flavins are the coenzymes that can serve as either one-electron or two-electron transfer reagents. Most of the biological oxidations or reductions that proceed in one-electron steps through free radicals are catalyzed by flavins. If the cell needs a two-electron oxidizing or reducing agent, it uses one of the pyridine nucleotide couples. But if it needs a one-electron oxidizing or reducing agent, it uses a flavin couple.

Flavins occur as two different derivatives. They are flavin mononucleotide, FMN, and flavin adenine dinucleotide, FAD. These are shown in the oxidized form along with the simple nutritionally important derivative called *riboflavin* (5.4). Riboflavin in your food is converted by the body to the catalytically active FMN or FAD. The flavins can be reduced by two electrons to the fully reduced forms (5.5).

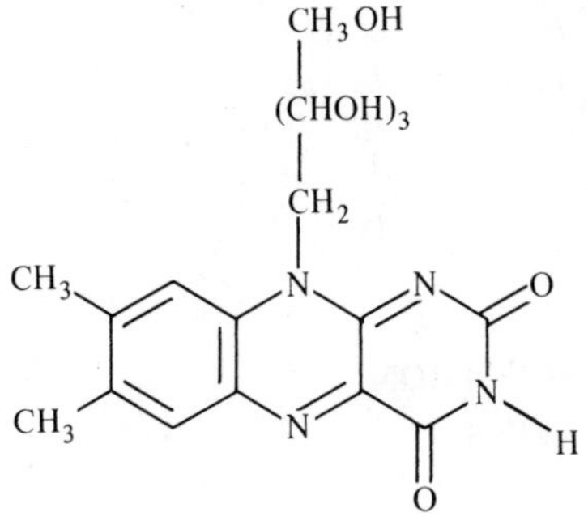

CH_3OH | $(CHOH)_3$ | CH_2 | N … CH_3, CH_3, N, N, O, N, H, O

Riboflavin

(5.4)

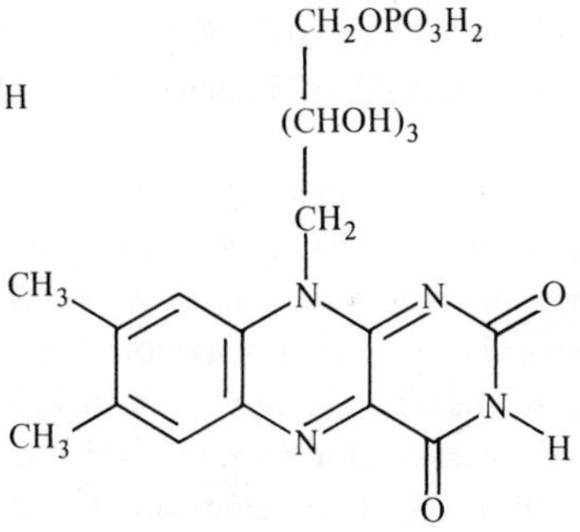

$CH_2OPO_3H_2$ | $(CHOH)_3$ | CH_2 | N … CH_3, CH_3, N, N, O, N, H, O

Flavin mononucleotide (FMN)

O O | $CH_2OPOP—O$ | $(HOHC)_3$ OH OH | CH_2 … CH_3, CH_3, N, N, O, NH, N, O; H_2C, O, OH OH; NH_2, N, N, N, N

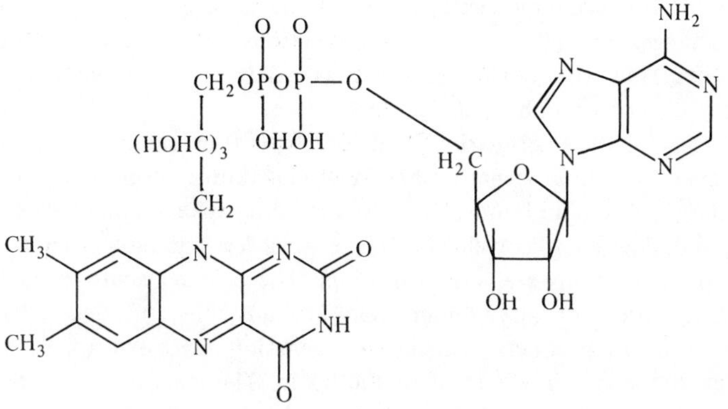

Flavin adonine dinucleotide (FAD)

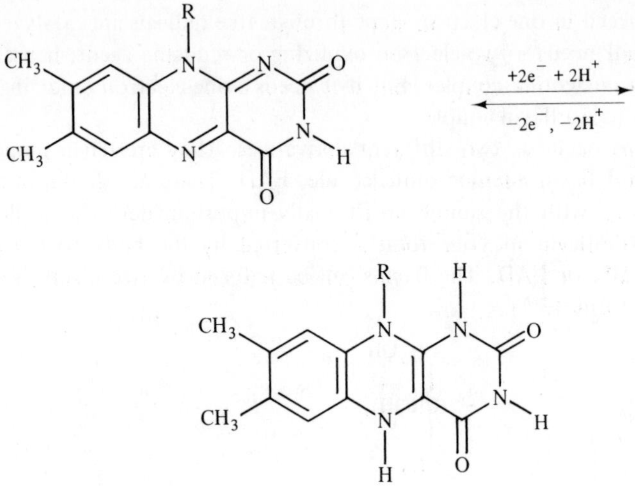

$$+2e^- + 2H^+ \rightleftharpoons -2e^-, -2H^+$$

(5.5)

5.2 ONE-ELECTRON AND TWO-ELECTRON REDOX REACTIONS OF FLAVINS

Flavins are able to catalyze free-radical reactions because they can be oxidized or reduced by one-electron steps. This is made possible by a stable-intermediate oxidation step between fully reduced and fully oxidized flavin. This stable intermediate is called a *semiquinone.* Flavin semiquinones are particularly stable because they have many extra resonance forms (5.6). The odd electron may be placed at several different positions in structural formulas indicating a large delocalization of the electron. The electron is essentially shared throughout the whole structure. The semiquinone intermediate may be further oxidized or reduced by one electron.

Often, two flavin molecules occur on the same enzyme in close proximity. This arrangement allows for redox reactions of one, two, three, or four electrons. Thus the flavins may act as catalysts to couple redox reactions between different numbers of electrons.

Flavins function primarily as dehydrogenation catalysts. In addition to catalyzing the dehydrogenation of saturated carbon-carbon single bonds to form carbon-carbon double bonds, they will dehydrogenate α-amino acids, α-hydroxy acids, and aldehydes, as well as catalyze some less common reactions. Some of the typical reactions are shown in (5.7). The α-imino acids formed from the dehydrogenation of an α-amino acid are unstable and decompose to give α-keto acids as products. The dehydrogenation reactions appear to proceed through the semiquinone form of the flavin. The reaction mixtures exhibit an electron spin resonance (ESR) signal. New electronic absorption bands appear in

(5.6)

(5.7)

$$\underset{\displaystyle |\atop OH}{RCHCOOH} + \text{Flavin} \rightleftharpoons \underset{\displaystyle \|\atop O}{RCCOOH} + \text{Flavin} \cdot H_2$$

(a)

$$H_2O + \underset{\displaystyle |\atop NH_2}{RCHCOOH} + \text{Flavin} \rightleftharpoons \underset{\displaystyle \|\atop O}{RCCOOH} + \text{Flavin} \cdot H_2 + NH_3$$

(b)

$$\begin{matrix} COOH \\ | \\ CH_2 \\ | \\ CH_2 \\ | \\ COOH \end{matrix} + \text{Flavin} \rightleftharpoons \begin{matrix} COOH \\ | \\ CH \\ \| \\ CH \\ | \\ COOH \end{matrix} + \text{Flavin} \cdot H_2$$

(c)

the 500–700 nm region characteristic of the semiquinone form of the flavins. All electrons are paired in the fully oxidized or fully reduced flavins so that these compounds do not exhibit ESR signals.

Many of these enzymes that require flavin as a cofactor contain either iron or molybdenum, or both. The function of these metal ions is not clearly known, but they could function by chelating with the flavin. Such bonding could allow another place for the odd electron to reside and thus stabilize the semiquinone form even more. The equilibrium between oxidized and reduced flavin to form the semiquinones is shifted toward the semiquinone on the enzyme. Metal ions are known to stabilize flavin semiquinones in solution.

Another explanation for the increased stability of the semiquinone on enzymes could be that the flavin semiquinone is stabilized by complexing with an aromatic amino acid on the protein. (Aromatic compounds are known to stabilize flavin semiquinones in solution.)

The flavins serve as cofactors in certain oxidative elimination reactions. The enzyme acyl dehydrogenase catalyzes the oxidation of butyryl coenzyme A to crotyl coenzyme A by FAD (5.8). Flavins can form semiquinone intermediates in oxidation, and this reaction shows an ESR signal indicative of a free radical.

(5.8)

$$CH_3CH_2CH_2\overset{\displaystyle O\atop \|}{C}SCoA + FAD \rightarrow CH_3CH{=}CH\overset{\displaystyle O\atop \|}{C}{-}S{-}CoA + FADH_2$$

(5.9) $$R\cdot + \;>CH{-}C\cdot< \;\rightarrow\; RH + \;>C{=}C<$$

Radicals are known to disproportionate, one being oxidized to form a double bond and the other being reduced (5.9). One source of radicals is the oxidation of an anion, as in the base catalyzed oxidation of 2-nitropropane (5.10). Oxidation (loss of electrons) is more favorable for an electron rich anion than for a neutral molecule.

(5.10)
$$H{-}C(CH_3)_2{-}NO_2 + OH^- \rightarrow {}^-{:}C(CH_3)_2{-}NO_2$$

$${}^-{:}C(CH_3)_2{-}NO_2 + R\cdot \rightarrow R{:}^- + \cdot C(CH_3)_2{-}NO_2$$

With these facts in mind, a plausible mechanism may be written for the eliminations catalyzed by acyl dehydrogenase (5.11). The increased acidity of the coenzyme A ester allows the formation of an anion which the flavin will oxidize to a radical (see Chapter 4, Section 4.2). Subsequent reaction with the resulting flavin radical will form the desired olefin.

$$RCH_2CH_2C(=O)SCoA \xrightarrow{-H^+} RCH_2\ddot{C}^-HC(=O)SCoA$$

(5.11) $$RCH_2\ddot{C}^-HC(=O)SCoA + FAD + H^+ \rightarrow RCH_2\dot{C}HC(=O)SCoA + FADH\cdot$$

$$RCH_2\dot{C}HC(=O)SCoA + FADH\cdot \rightarrow RCH_2{=}CHC(=O)SCoA + FADH_2$$

Even though the flavins act as one-electron oxidizing or reducing agents in two one-electron steps, they may be returned to the original state by one two-electron step. Thus, effectively they act as catalysts between reactants that donate or accept two electrons and reactants that accept or donate only one electron. A specific example is that between the two-electron NADH-NAD^+ couple and the one-electron cytochromes (5.12).

$$2\ \text{Cytochrome}\ Fe^{+++} + \text{(1-R-4H-pyridine-3-carboxamide, NADH)} \rightleftharpoons 2\ \text{Cytochrome}\ Fe^{++} + \text{(1-R-pyridinium-3-carboxamide, NAD}^+\text{)} + H^+ \tag{5.12}$$

5.3 FLAVIN DESIGN

Reactions between two-electron couples and one-electron couples are commonly slow without a catalyst. In the absence of a catalyst, two molecules of the one-electron reactant must react with one molecule of the two-electron reactant. Termolecular reactions are very rare. There are very few catalysts comparable to flavin needed by the organic chemist since the organic chemist can order a one-electron oxidant or reductant or a two-electron oxidant or reductant when needed. Exceptions are the developers in color photography. Developers serve as catalysts between the one-electron silver couple and the two-electron dye couple. The cell does not have the latitude of the organic chemist and must use for food and chemistry whatever is available.

In this section we will attempt to point out how an organic chemist would prepare a color photography developer with the implication that the cell must have faced similar problems in developing a catalyst to couple one-electron reactants with two-electron reactants. Parallel evolutionary processes are known in biology and have been well discussed. A similar evolutionary process that is seldom mentioned has taken place in the catalysis of chemical reactions. The evolution of catalysts for chemical reactions shows an interesting parallel between catalysts developed by chemists and those evolved in the cell. The parallel is, of course, not in time, for the evolution in organisms occurred over a much longer period of time than did the evolution of catalysts by organic chemists. Because of this parallel, it is possible to look at the trials and problems facing a chemist in developing a catalyst so as to gain a better appreciation for the catalysts found in nature.

Phenazines of various types are used in color photography in electron transport between the one-electron silver-silver ion couple and the two-electron oxidation-reduction dyes (5.13). This is possible because phenazines can undergo both the one-electron reactions required for the silver couple and the two-electron reaction required for the dye couple. These one-electron and two-electron reactions of phenazines are shown in (5.14). The one-electron reduction is quite easy because the free radical is stabilized by two aromatic rings (5.15).

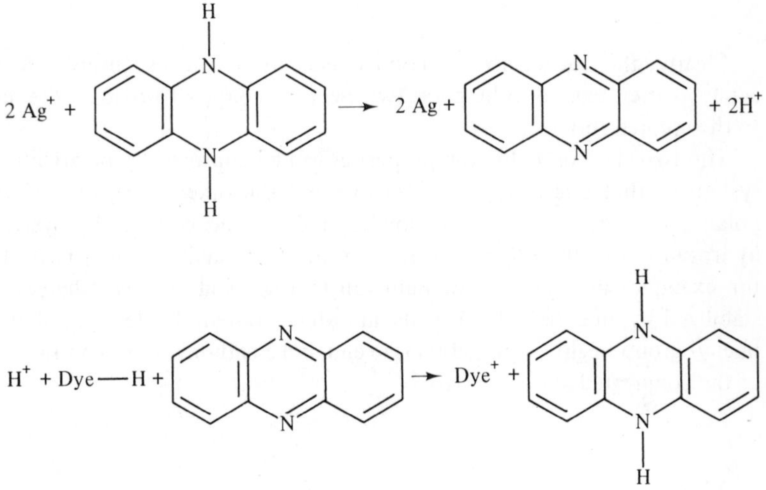

$2\,Ag^+$ + [dihydrophenazine] → $2\,Ag$ + [phenazine] + $2H^+$

H^+ + Dye—H + [phenazine] → Dye^+ + [dihydrophenazine]

(5.13)

[phenazine radical] $\underset{-H^+\,-e^-}{\overset{H^+\,e^-}{\rightleftharpoons}}$ [phenazine] $\underset{-2H^+\,-2e^-}{\overset{2H^+\,2e^-}{\rightleftharpoons}}$ [dihydrophenazine]

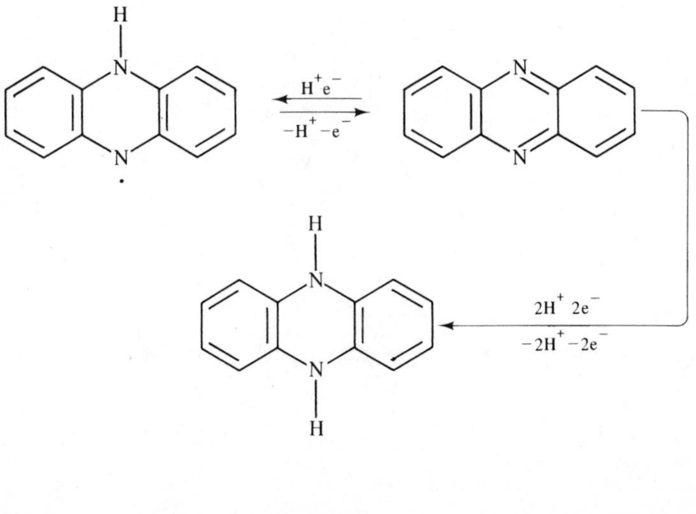

(5.14)

(5.15)

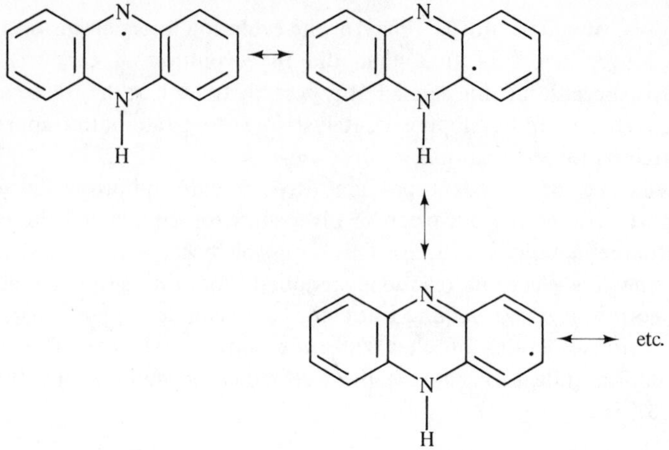

N N H ↔ N N H ↔ N N H ↔ etc.

Clearly, the phenazines are good one-electron couples. In order to be good catalysts they must also be good two-electron couples to oxidize the leuco dye to the colored form.

The two-electron reduction properties can be improved by polarization of the system so that one nitrogen is positive and the other is negative (5.16). This polarization enables a hydride ion to add to one end of the system and a hydrogen ion to the other. Positive ions are best stabilized by a phenyl ring as, for example, in phenyl carbonium ion (5.17a), and negative charges are best stabilized by an enol (5.17b). Thus, in order to obtain the desired polarization, a phenyl group might be placed on one end and a carbonyl group on the other end of the conjugated system (5.18).

(5.16)

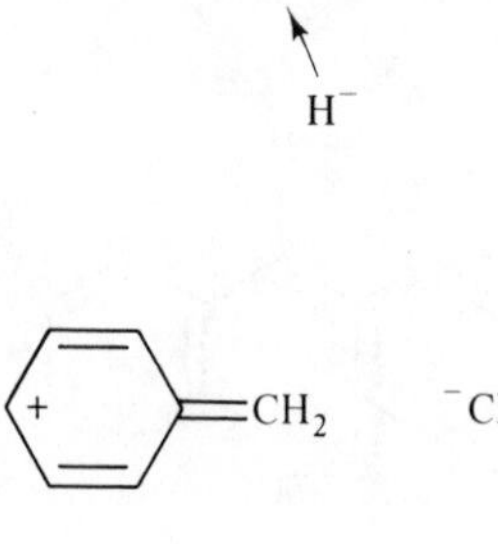

H^+ N N + H^-

(5.17)

$-CH_2^+ \leftrightarrow (+)=CH_2$ $\quad {}^-CH_2-C(=O)- \leftrightarrow CH_2=C(O^-)-$

(a) (b)

(5.18) $C_6H_5-\overset{+}{N}-CH=CH-\overset{-}{N}-\overset{O}{\overset{\|}{C}}-R$

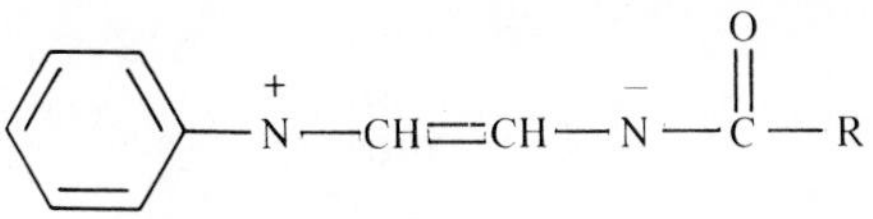

Because maximum resonance is obtained when the system is planar, the catalyst could be improved by adding some bridges to prevent rotation (5.19). However, the great advantage of the phenazines as catalysts is that the aromaticity of two rings helps to stabilize the intermediate free radical. Compound (5.19) has only one fully aromatic ring. The ring on the right may be made aromatic also by adding either a double bond or a group that can easily form a double bond by enolization (such as an amide group). Thus an ideal catalyst would have the general form of (5.20).

(5.19)

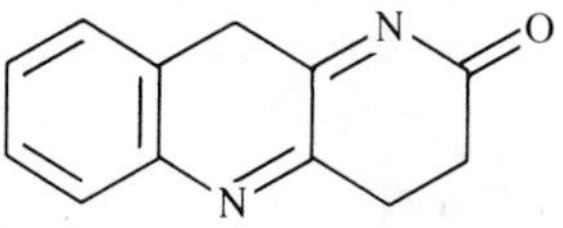

(5.20)

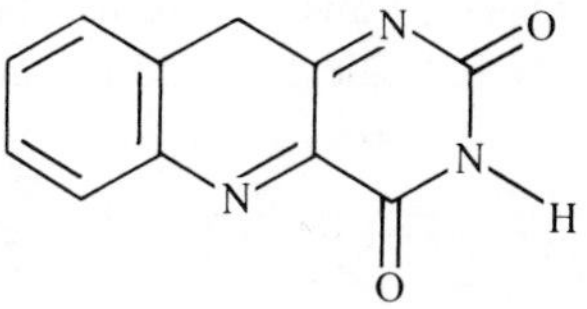

There is a problem of solubility. These catalysts must operate in an aqueous solution of protein (film gelatin). A sugar or other polar group could be added to increase the water solubility of the catalyst, but if such a group is added, it should be added to the upper side of (5.20) because the nitrogen on the lower side of the structure must react with the large dye molecule. Any solubilizing group on the lower side would sterically interfere with the hydride transfer. The most logical solution to the solubility problem would be to add a polyhydroxyl group to the upper side of the left or center ring, as shown in (5.21).

(5.21) $CH_2OH-(CHOH)_3-CH_2-$

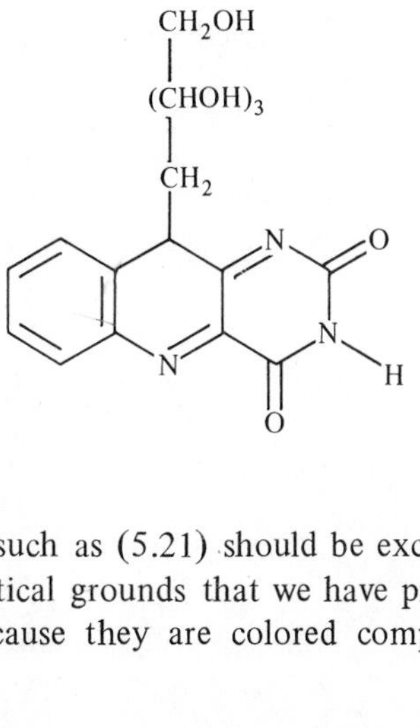

Although compounds such as (5.21) should be excellent catalysts for color developing on the theoretical grounds that we have presented, they cannot be used in photography—because they are colored compounds themselves, they

(5.22)

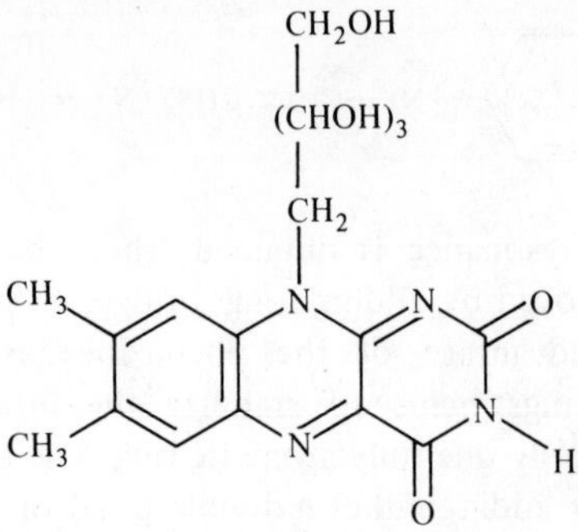

are unsuitable. The flavin catalyst (5.22) that the cell uses is remarkably similar to that predicted to be well suited for use as a color developer. Unlike the film maker, however, we can tolerate color that is more than skin deep. Flavins are the biological solution to coupling one- and two-electron redox reactions.

5.4 THIAMINE—A BIOLOGICAL CYANIDE ION

Thiamine is a coenzyme that participates in reactions forming and breaking carbon-carbon bonds. Forward and reverse condensation of coenzyme A esters are used to form or cleave bonds when activated groups are beta to one another (Chapter 4). Thiamine, like cyanide ion for the organic chemist, can be used in cases where the activated carbons are alpha to one another. The structure of thiamine is given in (5.23a). When thiamine acts as a cofactor on an enzyme, it occurs as the pyrophosphate derivative (5.23b).

Thiamine

(a)

(5.23)

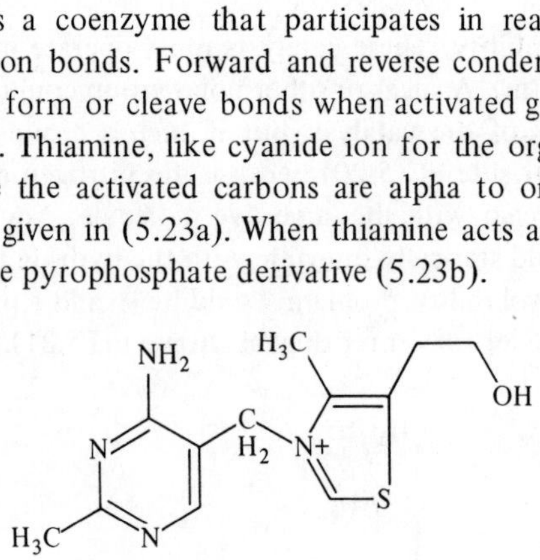

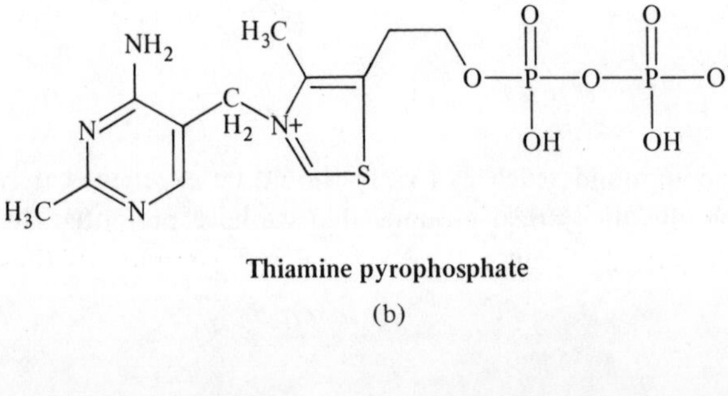

Thiamine pyrophosphate

(b)

There are four general functions of thiamine in the body. Thiamine acts as a cofactor to enzymes that catalyze the decarboxylation of α-keto acids (5.24). For example, the enzymes that catalyze the decarboxylation of pyruvic acid and α-ketoglutaric acid require thiamine as a cofactor.

$$\text{(5.24)}\quad \mathrm{R\overset{O}{\overset{\|}{C}}COOH} \rightarrow \mathrm{RCHO} + CO_2 \qquad \text{(5.25)}\quad \mathrm{RCHO} + \mathrm{R'CHO} \rightarrow \mathrm{R\overset{O}{\overset{\|}{C}}\overset{OH}{\overset{|}{C}}HR'}$$

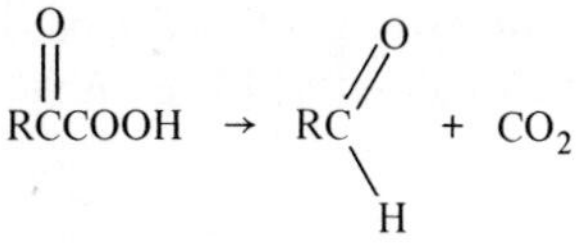

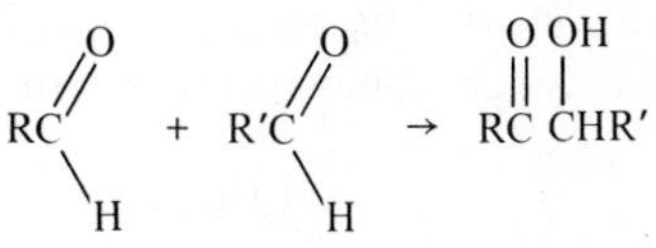

Thiamine also catalyzes benzoin-type condensations (5.25). Several examples of this type of reaction in metabolism are found in the pentose pathway. In the net reaction, pentose phosphate reacts to form a heptose phosphate and a triose phosphate. The pathway is given in (5.26). In this conversion, the xylulose-5-phosphate is cleaved by a reverse benzoin-type condensation to glycol

(5.26)

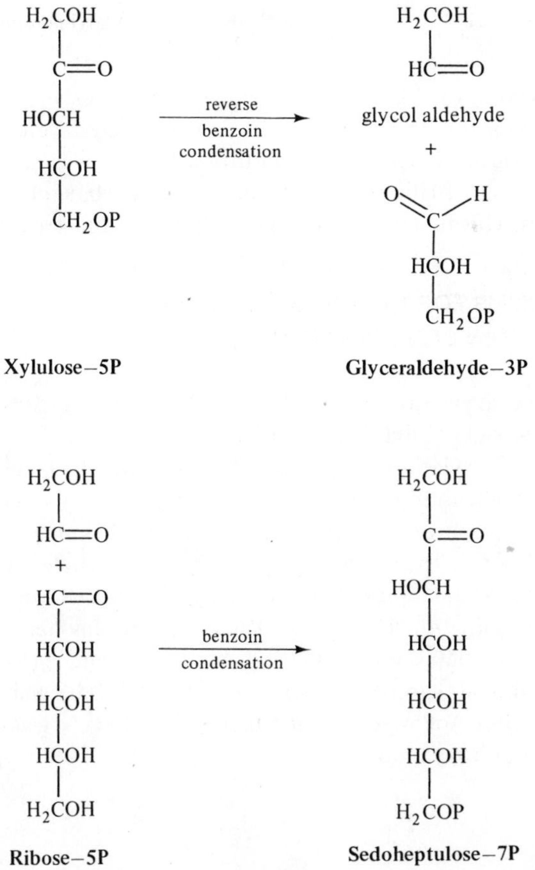

Xylulose–5P (H_2COH–C=O–HOCH–HCOH–CH_2OP) $\xrightarrow[\text{benzoin condensation}]{\text{reverse}}$ glycol aldehyde (H_2COH–HC=O) + Glyceraldehyde–3P (O=C(H)–HCOH–CH_2OP)

glycol aldehyde (H_2COH–HC=O) + Ribose–5P (HC=O–HCOH–HCOH–HCOH–H_2COH) $\xrightarrow[\text{condensation}]{\text{benzoin}}$ Sedoheptulose–7P (H_2COH–C=O–HOCH–HCOH–HCOH–HCOH–H_2COP)

aldehyde and glyceraldehyde-3-phosphate. The glycol aldehyde is then condensed with ribulose-5-phosphate by a benzoin condensation.

A third fascinating type of reaction catalyzed by enzymes containing thiamine pyrophosphate as a cofactor is the reaction of xylulose-5-phosphate with phosphate ion to form glyceraldehyde-3-phosphate and acetyl phosphate (5.27). In this reaction, a high-energy phosphate ester is formed from a sugar. Acetyl phosphate has a free energy of hydrolysis of about − 14 kcal/mole.

$$\begin{array}{c} CH_2OH \\ | \\ C{=}O \\ | \\ HOCH \\ | \\ HCOH \\ | \\ H_2COP \end{array} + P_i \longrightarrow \begin{array}{c} CH{=}O \\ | \\ CHOH \\ | \\ CH_2OP \end{array} + \begin{array}{c} CH_3 \\ | \\ O{=}C{-}OP \end{array} \qquad (5.27)$$

Xylulose-5P **Glyceraldehyde-3P** **Acetyl phosphate**

A fourth function of thiamine is its action in nerve tissue. Nerve tissue has a high concentration of thiamine. Thiamine deficiencies were first observed clinically as nervous disorders and the nutritional factor was named aneurin, or the antineuritic factor. However, it is still not clear what role the thiamine is playing in nerves. This function of thiamine will not be discussed in this book.

5.5 MECHANISMS FOR DECARBOXYLATIONS AND BENZOIN CONDENSATIONS

When the organic chemist wishes to catalyze either (5.24) or (5.25), he can use a cyanide salt. Although cyanide is not commonly used in decarboxylations, it will catalyze the decarboxylation of an α-keto acid. Cyanide is the classical catalyst for the benzoin condensation.

The decarboxylation of pyruvic acid catalyzed by thiamine was presented in Chapter 4, Section 4.3. The mechanism of catalysis by cyanide is given in (5.28). In the first step, cyanide adds to the carbonyl group to form a cyanohydrin (5.28a). The cyanohydrin is decarboxylated more readily than the keto acid because the cyano group stabilizes the carbanion intermediate (formed upon the loss of carbon dioxide) through resonance forms [(5.28b) and (5.28c)]. The addition of a proton followed by the elimination of HCN gives the aldehyde resulting from decarboxylation.

(5.28)

$$R{-}\overset{\overset{O}{\|}}{C}{-}CO_2^- \quad + \quad CN^- \quad + \quad H^+ \;\rightleftharpoons\; R{-}C(OH)(CO_2^-){-}C{\equiv}\ddot{N}$$

(a)

$$\Big\downarrow -CO_2$$

$$R{-}\ddot{C}^-(OH){-}C{\equiv}N{:} \;\longleftrightarrow\; R{-}C(OH){=}C{=}\ddot{N}{:}^-$$

(b) (c)

$$\Big\updownarrow H^+$$

$$R{-}C(OH)(H){-}C{\equiv}N \;\overset{-\,HCN}{\rightleftharpoons}\; R{-}C(=O){-}H$$

A similar mechanism occurs in the benzoin condensation (5.29). The cyanide ion adds to benzaldehyde to produce benzaldehyde cyanohydrin (5.29a). The cyano group and the hydroxyl group strongly withdraw electrons from carbon, weakening the carbon-hydrogen bond. This electron withdrawal, together with the resonance stabilization of the resulting carbanion (5.29b), makes this hydrogen acidic. The carbanion (5.29b) can in turn condense with another molecule of benzaldehyde to give the anion of the cyanohydrin of benzoin (5.29c). Loss of cyanide ion gives the final product, benzoin.

$$R-CHO + CN^- + H^+ \rightleftharpoons R-CH(OH)-CN$$

(a)

$$\xrightleftharpoons{-H^+}$$

$$R-\ddot{C}(OH)-C\equiv N: \longleftrightarrow R-C(OH)=C=\ddot{N}:$$

(b)

(5.29)

$$+ \quad R-CHO$$

$$\rightleftharpoons$$

$$R-C(OH)(CN)-C(O^-)(H)-R \rightleftharpoons R-C(O^-)(CN)-C(OH)(H)-R$$

(c)

$$\rightleftharpoons$$

$$R-C(=O)-CH(OH)-R \quad + \quad CN^-$$

5.6 THIAMINE DESIGN

The chemical suitability of thiamine as a catalyst will be discussed in this section. As was done with flavins, it will be argued that thiamine has the optimum chemical structure for its biological role. In a sense, this argument may be termed chemical adaptation. Thiamine would then be regarded as the result of chemical evolution.

In the discussion of flavin design, a solution to a chemical problem was presented, then improved upon. That is not the case in this section. Thiamine is compared with hypothetical catalysts, compounds that for one reason or another fail to have the ability to catalyze the desired reactions. By finding how chemical "competitors" are deficient, we can identify the way that thiamine is "adapted" to its role.

The structural features that enable cyanide to catalyze the decarboxylation of α-keto acids and the benzoin condensations have been identified. Cyanide is a good nucleophile. Once added to a carbonyl group it can stabilize a negative charge at the alpha carbon atom by induction and resonance. Finally, it is a good leaving group. However it is toxic to living hosts because of its affinity for iron (see Chapter 7). We will look for a nontoxic alternative catalyst having these same features.

The primary function of cyanide ion in (5.24) and (5.25) is to stabilize the carbanion intermediates. By resonance, the nitrogen atom of the cyano group acts as an electron sink for the negative charge. A species in which the nitrogen was positive rather than neutral should be an even better electron sink.

A possible catalyst with such a positive nitrogen would be an alkylated Schiff base as, for example, the methylated imine formed between acetaldehyde and aniline (5.30). Although this compound has a positive nitrogen atom, it is not a

(5.30) $CH_3CH{=}\overset{+}{N}(CH_3)-C_6H_5$

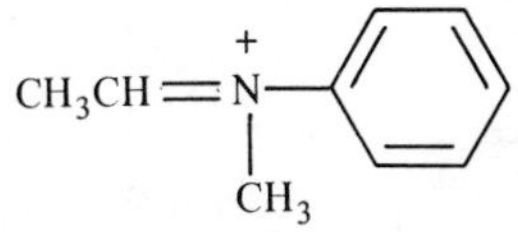

nucleophile and does not add to the carbonyl group of benzaldehyde. Hydrogen cyanide itself is not a nucleophile, but its conjugate base, cyanide ion, is. The hydrogen attached to the imino carbon atom would be expected to be acidic, however, and the resulting carbanion (5.31) should be a good nucleophile. There are two reasons to believe this proton would be acidic. First, it is adjacent to a strongly electron-withdrawing positive nitrogen, and second, a carbene canonical form (5.31b) can contribute to the resonance stabilization of the zwitterion. A carbene is a neutral, divalent carbon species with six electrons instead of the usual eight about carbon. The simplest carbene is methylene ($:CH_2$).

$$(5.31)\qquad \underset{(a)}{CH_3{-}\ddot{C}^{-}{=}\overset{+}{N}(CH_3){-}C_6H_5} \longleftrightarrow \underset{(b)}{CH_3{-}\ddot{C}{-}\ddot{N}(CH_3){-}C_6H_5}$$

When (5.30) is treated with base, it polymerizes. The most acidic hydrogen is beta to the positive nitrogen, and an enamine isoelectronic to an enolate ion is formed by removal of a proton (5.32). This adds to another molecule of the Schiff base (5.33), and the product can react further.

$$(5.32)\qquad CH_3CH{=}\overset{+}{N}(CH_3){-}C_6H_5 \xrightarrow{-H^+} {}^{-}\ddot{C}H_2CH{=}\overset{+}{N}(CH_3){-}C_6H_5 \longleftrightarrow CH_2{=}CH{-}\ddot{N}(CH_3){-}C_6H_5$$

$$(5.33)\qquad CH_2{=}CH{-}\ddot{N}(CH_3){-}C_6H_5 + CH_3{-}CH{=}\overset{+}{N}(CH_3){-}C_6H_5 \longrightarrow CH_3CH(\ddot{N}(CH_3)C_6H_5){-}CH_2{-}CH{=}\overset{+}{N}(CH_3){-}C_6H_5$$

In designing a catalyst, therefore, we cannot have hydrogen bound to a carbon atom attached to the imino carbon. This feature can be avoided by using benzaldehyde in place of acetaldehyde to synthesize (5.34). But this ion has other problems. It reacts very rapidly with water to give benzaldehyde and N-methylanaline. Imine hydrolysis is acid catalyzed (5.35), and (5.34) corresponds electronically to the highly reactive protonated form of the Schiff base. A catalyst is of no use in an aqueous system if it is destroyed immediately by water.

(5.34)

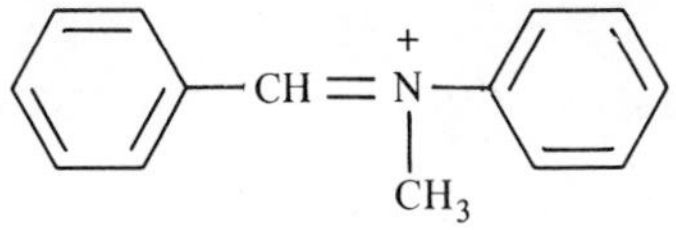

$$\mathrm{C_6H_5{-}CH{=}\overset{+}{N}(CH_3){-}C_6H_5}$$

(5.35)

$$\mathrm{R_1CH{=}\ddot{N}{-}R_2 \underset{-H^+}{\overset{H^+}{\rightleftharpoons}} R_1CH{=}\overset{+}{N}(H){-}R_2 \xrightarrow{H_2O} R_1CH{=}O + H_2NR_2}$$

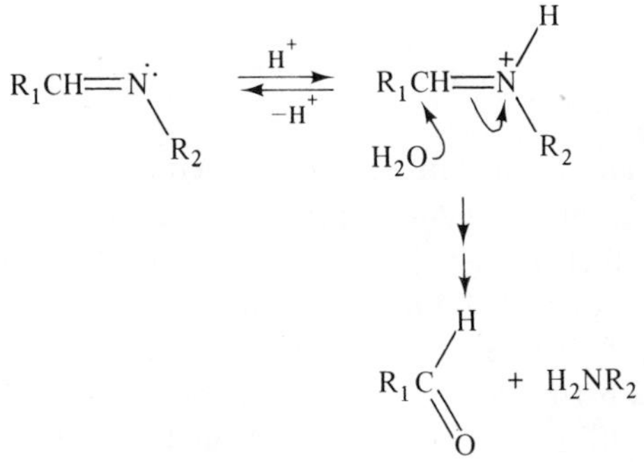

Methylated thioamides are known to be stable in water, presumably because the sulfur shares part of the positive charge with the nitrogen. The next trial catalyst was (5.36). Compound (5.36) was found to have other difficulties, however. This compound is stable in water for a long time, but it is not acidic. When (5.36) is dissolved in D_2O for three hours at pH 7, there is no deuterium incorporation. If the proton were acidic, the carbanion would have been formed. This anion would have picked up a deuterium ion from D_2O. Another failure in the search for a catalyst, but yet another lesson learned.

(5.36)

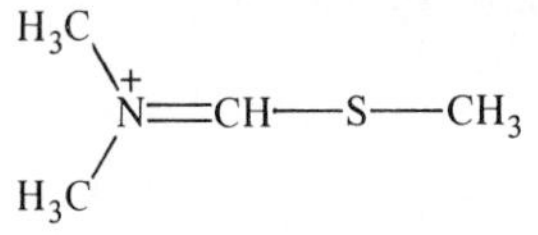

$$\mathrm{(H_3C)_2\overset{+}{N}{=}CH{-}S{-}CH_3}$$

Consider (5.30) and (5.31). The product of the reaction has less double bond character than the reactant, so that any strain on the double bond should cause the reaction to proceed to the right and thus make the proton more acidic. Also consider the structure of the carbene (5.31b). There are two types of carbenes–one in which the two electrons are paired, called a *singlet,* and the other in which the two electrons are unpaired, called a *triplet.* Since both (5.31a) and (5.31b) are contributing to the structure of the product, the carbene (5.31b) must be a singlet, because resonance is not allowed between forms of different spin multiplicity. The most stable structure of a singlet carbene has an angle around the carbene carbon of about 104°. Therefore, if we reduce the angle below 120° around the carbon, we will strain the reactant and tend to stabilize the product (5.31). This can be done by placing the carbon in a five-membered ring as shown in (5.37b). Compound (5.37a) is (5.36) drawn so as

(5.37)

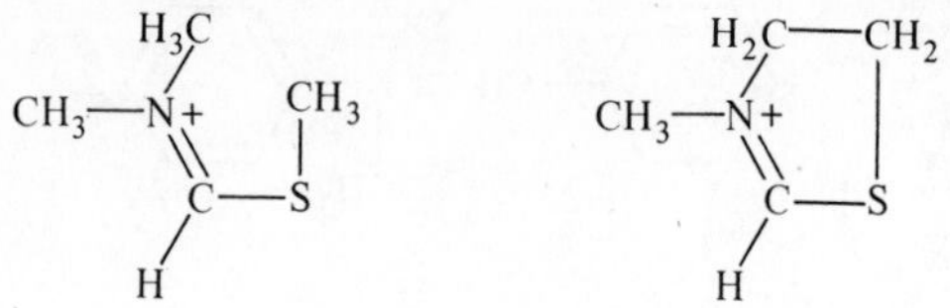

$$\mathrm{CH_3{-}\overset{H_3C}{N^+}{=}\underset{H}{C}{-}S{-}CH_3}\qquad\qquad \text{(cyclic: } \mathrm{H_2C{-}CH_2}, \mathrm{CH_3{-}N^+{=}C(H){-}S}\text{)}$$

(a) (b)

to compare it more easily with (5.37b). Experimentally, (5.37a) was not acidic, but (5.37b) was found to be highly acidic in agreement with predictions.

However, another problem arose. Compound (5.37b) is extremely unstable in water because the double bond is strained, and it again corresponds to a protonated Schiff base. Some resonance stability can be added by having two nitrogens with a partial charge on each nitrogen instead of on a nitrogen and a sulfur. Compound (5.38) was prepared and found to have both of the requirements for a good catalyst. The compound is relatively stable in water, yet the proton is acidic. The acidity was measured by the exchange of the proton for a deuterium in D_2O. The exchange is base catalyzed so it had to be measured at slightly acidic pH conditions to be experimentally feasible. The time for 50% exchange was 10 minutes at pH 5.6 in deuterium oxide. This can be compared with similar exchange rates for tetramethylammonium iodide which would produce a carbanion stabilized only by the positive charge on nitrogen. The rate of exchange in deuterium oxide for tetramethylammonium iodide is much slower. In boiling 0.1 *N* base, only 0.1% exchange is obtained in 15 days. There is a remarkable difference in exchange rates between these compounds. The greater acidity of (5.38) must be ascribed to the carbene structure, which helps stabilize the carbanion intermediate.

(5.38)

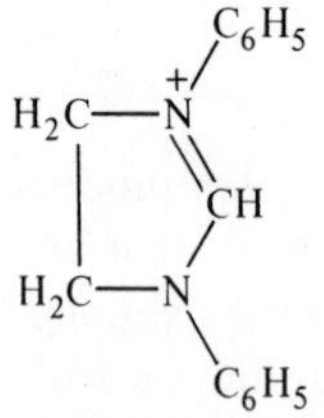

$$\text{ring: } \mathrm{H_2C{-}N^+(C_6H_5){=}CH{-}N(C_6H_5){-}H_2C}$$

With these two problems solved, acidity and stability, we are ready to try (5.38) as a catalyst. Compound (5.38) has no activity as a catalyst. It reacted with benzaldehyde but gave (5.39) instead of benzoin. There must be further considerations in designing a catalyst. A logical mechanism that accounts for the formation of (5.39) is given in (5.40). The nucleophile is formed and adds to benzaldehyde. The removal of the proton from the carbinol carbon gives a stable carbanion (5.40b ↔ c). This carbanion is stabilized by the positive nitrogen, too much so to react further. Structure (5.40c) is a major contributing structure to

(5.39)

H_5C_6 O
H_2C—N C—C_6H_5
C
H_2C—N H
C_6H_5

(5.40)

C_6H_5 N$^+$ H N C_6H_5 + $:B(-H^+)$ ⇌ C_6H_5 N$^+$:$^-$ N C_6H_5 + C_6H_5C(=O)H

H^+

C_6H_5 N$^+$ OH C—C_6H_5 H N C_6H_5

$:B(-H^+)$ (a)

C_6H_5 N OH =C—C_6H_5 N C_6H_5 ⟷ C_6H_5 N$^+$ OH C—C_6H_5 N C_6H_5

(b) (c)

ketonization

C_6H_5 N O C—C_6H_5 H N C_6H_5

the resonance hybrid. There is not enough contribution of (5.40b) to add to another benzaldehyde molecule. This enol (5.40b ↔ c) ketonizes to give the observed product, (5.39).

As a potential catalyst, (5.39) was not a complete failure. It did promote the acidity of hydrogen at the alpha position. A compound that could do these tasks, but which would not neutralize the carbanion to such a large extent, would be an excellent candidate for a catalyst. In the resonance, (5.40b ↔ c), the double bond is endocyclic in the zwitterion canonical structure, and exocyclic in the enolic structure. In order to keep a greater negative charge density on carbon, the hybrid must have a greater contribution from structures with double bond character in the ring. This change should be facilitated by increased resonance in the ring.

A good catalyst might be an *N*-methylpyridinium cation (5.41a ↔ b). This has a positive nitrogen, a ring, and increased resonance in the ring. Another problem arises, however. The hydrogen attached to the alpha positions of the aromatic ring are not acidic enough for reaction. The potential nucleophile (5.41c ↔ d) is not formed. Apparently the pyridinium salt is too stable. Compared to (5.38), the extra aromatic resonance stabilizes the pyridinium ion more than the resonance involving a carbene (5.41c ↔ d) stabilizes the conjugate base.

(5.41)

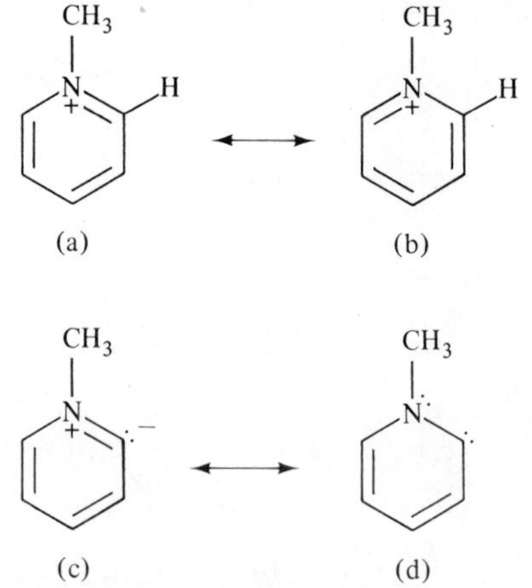

(a) (b)

(c) (d)

The solution to the problem has been bracketed in these experiments. With too much resonance in the ring, as in *N*-methylpyridinium salts, no nucleophile forms. In the other extreme case, with too little resonance in the ring, the nucleophile forms but the intermediate carbanion ketonizes.

The solution to the problem is to put partial aromatic character into the ring of the catalyst. A partial double bond would provide a compound intermediate between the pyridinium ion (5.41) and the imidazolinium ion (5.38). Sulfur often acts as a partial double bond in conjugated systems. The sulfur atom does conduct electrons of π systems, but does not do so well as a double bond.

Replacement of a double bond in a Kekulé structure of a pyridinium ion gives a thiazolium ion (5.42). Thiazolium salts have been synthesized and have been shown to be excellent catalysts for the formation of benzoin from benzaldehyde. The catalyst found in nature, thiamine, is a thiazolium salt.

(5.42)

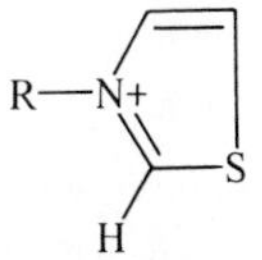

R—N+ S H

Another way in which the sulfur atom may promote catalysis by thiazolium salts involves valence shell expansion. The carbanion intermediate (5.43) may be stabilized by (5.43c), which places negative charge density on sulfur. The quantitative importance of this resonance is not known.

(5.43)

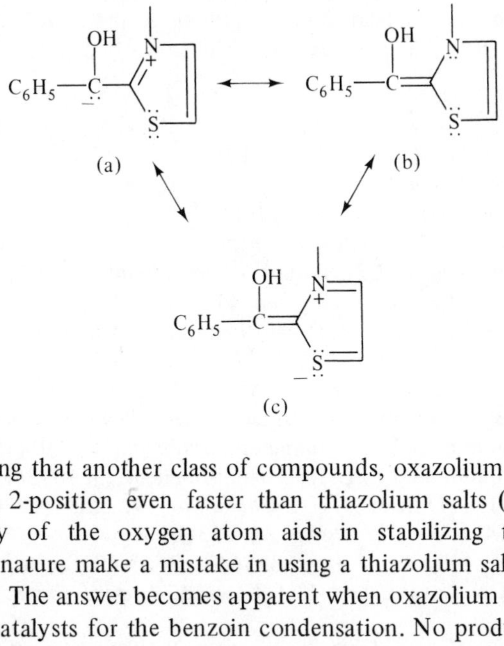

OH N+ C_6H_5—C S (a) ⟷ OH N C_6H_5—C S (b)

OH N+ C_6H_5—C S (c)

It is intriguing that another class of compounds, oxazolium salts, exchanges protons at the 2-position even faster than thiazolium salts (5.44). The high electronegativity of the oxygen atom aids in stabilizing the nucleophilic carbanion. Did nature make a mistake in using a thiazolium salt rather than an oxazolium salt? The answer becomes apparent when oxazolium salts are studied. They are not catalysts for the benzoin condensation. No products are formed, not even adducts comparable to (5.39). The explanation of these results is that the oxazolium salt is too stable to add to benzaldehyde. The anion must be stable enough to form, but not so stable that it is unreactive.

(5.44)

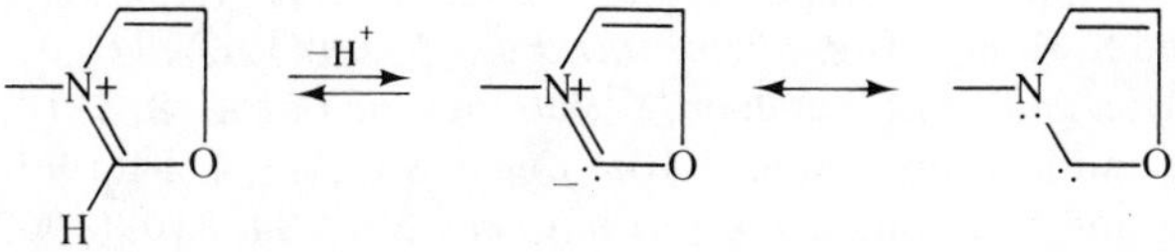

—N+ O H $\overset{-H^+}{\rightleftharpoons}$ —N+ O ⟷ —N O

Thiamine is a beautifully designed catalyst for the benzoin condensation. It represents a balance between many problems. Considerable intelligence was used in the design of thiamine.

REFERENCES

Flavins and Flavin Models

H. Beinert and P. Hemmerich, *Biochem. Biophys. Res. Comm.*, **18**, 212(1965).
R. D. Draper and L. L. Ingraham, *Arch. Biochem. Biophys.*, **125**, 802(1968).
R. D. Draper and L. L. Ingraham, *Arch. Biochem. Biophys.*, **139**, 265(1970).
P. Hemmerich, *Helv. Chem. Acta*, **47**, 464(1964).
H. R. Mahler, *Adv. Enz.*, **17**, 233(1956).
A. H. Neims and L. Hellerman, *Ann. Rev. Biochem.*, **39**, 867(1970).
D. J. T. Porter, J. G. Voet, and H. J. Bright, *J. Biol. Chem.*, **247**, 1951(1972).
P. Strittmatter, *Ann. Rev. Biochem.*, **35**, 125(1966).
C. H. Suelter and D. E. Melzler, *Biochem. Biophys. Acta*, **44**, 23(1960).
D. Wellner, *Ann. Rev. Biochem.* **36**, 669(1967).

Thiamine

R. Breslow, *Chem. Ind.* (London), **R28**(1956).
R. Breslow, *J. Am. Chem. Soc.*, **79**, 1762(1957).
R. Breslow, *J. Am. Chem. Soc.*, **80**, 3719(1958).
J. Crosby and G. E. Leinhard, *J. Am. Chem. Soc.*, **92**, 5707(1970).
J. Crosby, R. Stone and G. E. Leinhard, *J. Am. Chem. Soc.*, **92**, 2891(1970).
V. Franzen and L. Fikentscher, *Annalen*, **613**, 1(1958).
K. Fry, L. L. Ingraham, and F. H. Westheimer, *J. Am. Chem. Soc.*, **79**, 5225 (1957).
P. Haake and L. P. Bauscher, *J. Phys. Chem.*, **72**, 2213(1968).
P. Haake, L. P. Bauscher, and W. B. Miller, *J. Am. Chem. Soc.*, **91**, 1113(1969).
W. Hafferl, R. Lundin, and L. L. Ingraham, *Biochem.*, **2**, 1298(1963).
W. Hafferl, R. Lundin, and L. L. Ingraham, *Biochem.*, **3**, 1072(1964).
D. S. Kemp and J. T. O'Brien, *J. Am. Chem. Soc.*, **92**, 2554(1970).
L. O. Krampitz, *Ann. Rev. Biochem.*, **38**, 213(1968).
G. E. Leinhard, *J. Am. Chem. Soc.*, **88**, 5642(1966).
D. M. Lemal, R. A. Lovald, and K. I. Kawano, *J. Am. Chem. Soc.*, **86**, 2518 (1964).
J. J. Mieyal, R. G. Votaw, L. O. Krampitz, and H. Z. Sable, *Biochem. Biophys. Acta.*, **141**, 205(1967).
C. P. Nash, C. W. Olson, F. G. White, and L. L. Ingraham, *J. Am. Chem. Soc.*, **83**, 4106(1961).
R. A. Olofson and J. M. Landesberg, *J. Am. Chem. Soc.*, **88**, 4263(1966).
H. Quast and S. Hunig, *Ang. Chem. (Int. Ed.)*, **3**, 800(1964).
H. C. Sorenson and L. L. Ingraham, *J. Heterocyclic Chem.*, **8**, 551(1971)
F. G. White and L. L. Ingraham, *J. Am. Chem. Soc.*, **82**, 4114(1960).
F. G. White, and L. L. Ingraham, *J. Am. Chem. Soc.*, **84**, 3109(1962).

6 CATALYSIS BY PROTEIN FUNCTIONAL GROUPS

6.1 INTRODUCTION

There are many enzymatic reactions that do not involve a coenzyme or cofactor. The nucleophilic basic and acidic groups on the protein function as catalysts, sometimes with the aid of other nonprotein molecules. These reaction mechanisms are best exemplified by certain hydrolyses, eliminations, and rearrangements. Elimination reactions of amino acids, however, are catalyzed by enzymes containing pyridoxal as a cofactor (see Chapter 3). We will first discuss hydrolytic and esterification reactions and then examine elimination reactions.

6.2 HYDROLYTIC REACTIONS

Hydrolytic reactions are commonly catalyzed by enzymes that do not have cofactors. The catalyses are entirely performed by groups on the protein. We will present the normal acid and base catalyzed hydrolyses of esters and then discuss how other groups can function as catalysts. When ethyl benzoate labeled with O^{18} in the carbonyl oxygen is hydrolyzed to benzoic acid and ethanol in either acidic or basic solution, some of the O^{18} is lost to solvent. These observations are consistent with mechanisms which proceed through tetrahedral intermediates. The base-catalyzed ester hydrolysis proceeds by the attack of a hydroxide ion on the carbonyl of the ester followed by the decomposition of the intermediate to give the acid and alcoholate ion (6.1). Acid catalyzed ester hydrolysis, shown in (6.2), proceeds by the attack of water on the protonated ester.

$$\text{(6.1)}\quad \text{R—C}(=\overset{*}{\text{O}})\text{—OR}' + \text{HO}^- \rightleftharpoons \text{R—C}(\overset{*}{\text{O}}{}^-)(\text{OH})\text{—OR}' \longrightarrow \text{R—C}(=\overset{*}{\text{O}})\text{—OH} + \text{OR}^-$$

$$\text{R—C}(\text{O}^-)(\overset{*}{\text{O}}\text{H})\text{—OR}' \rightleftharpoons \text{R—C}(=\text{O})\text{—OR} + {}^-\overset{*}{\text{O}}\text{H}$$

$$\text{R—C}(\text{O}^-)(\overset{*}{\text{O}}\text{H})\text{—OR}' \longrightarrow \text{R—C}(=\text{O})\text{—}\overset{*}{\text{O}}\text{H} + \text{OR}^-$$

$$\text{(6.2)}\quad \text{R—C}(=\overset{*}{\text{O}}{}^+\text{H})\text{—OR}' + \text{OH}_2 \rightleftharpoons \text{R—C}(\overset{*}{\text{O}}\text{H})(\text{O}^+\text{H}_2)\text{—OR}' \rightleftharpoons \text{R—C}(\overset{*}{\text{O}}\text{H})(\text{OH})\text{—O}^+\text{(H)R} \longrightarrow \text{R—}\overset{*}{\text{C}}\text{OOH}_2^+ + \text{ROH}$$

$$\text{R—C}(\overset{*}{\text{O}}\text{H})(\text{O}^+\text{H}_2)\text{—OR}' \rightleftharpoons \text{R—C}(+\overset{*}{\text{O}}\text{H}_2)(\text{OH})\text{—OR}' \rightleftharpoons \text{R—C}(+\overset{*}{\text{O}}\text{H}_2)(=\text{O}^+\text{H})\text{—OR}'$$

Thus the exchange data during the hydrolysis of an ester require the existence of a tetrahedral intermediate. A tetrahedral intermediate is also attractive on theoretical arguments. The ester has bonds with sp^2 hybridization. The greater the s-character of a hybrid orbital, the shorter and stronger the bond. The tetrahedral intermediate uses sp^3 hybridized orbitals in bonding. In the tetrahedral intermediate, the bond force constant is lower; the carbon-oxygen bond is more easily stretched. Since less energy is required to reach the transition state for bond cleavage, the reaction is more rapid than for a trigonal species.

The hydrolysis of esters by the above mechanisms takes place in either acidic or basic solution. However, as discussed in Chapter 1, the fascinating property of enzymes is that they are able to catalyze these reactions at pH 7, although some of the esterases such as pepsin operate at much lower pH values. One of the important properties of enzymes is that they can have both acidic and basic groups. This allows the mechanism to be a combination of acid and base catalysis. Another important factor is that these groups can be in the proper spatial arrangement for catalysis. Since hydroxide ions and hydronium ions are not available to the enzymes at pH 7, it is of great interest to know whether acidic and basic groups that are available to the enzymes can catalyze ester hydrolysis.

The groups that are particularly important in hydrolytic reactions are the carboxyl group, the thiol group and the imidazole group. There has been a considerable amount of work on the mechanism of hydrolytic reactions catalyzed by these groups. Model systems can be used to explore catalysis–they cannot be used to determine which of these mechanisms is used by the enzymes.

A carboxylate anion can act as a base to catalyze the hydrolysis of an ester. For example, acetate ion will catalyze the hydrolysis of an ester. The hydrolysis of 2,4-dinitrophenyl benzoate, when catalyzed by O^{18} labeled acetate ion, produces O^{18} labeled benzoate ion. This result is consistent with a reaction mechanism proceeding through an anhydride (6.3). The acetate ion attacks to form a mixed anhydride of acetic and benzoic acid. Water may attack this anhydride on the carbonyl site of either the benzoyl group or the acetyl group. Attack on the acetyl group would give labeled benzoate ion.

(6.3)

$$CH_3-C(=O^*)-O^{*-} \dashrightarrow C_6H_5-C(=O)-O-C_6H_3(NO_2)_2 \longrightarrow CH_3-C(=O^*)-O^*-C(=O)-C_6H_5$$

$$\downarrow$$

$$CH_3-COO^{*-} + C_6H_5-CO_2^{*-}$$

Several models of esterases have demonstrated that the reaction is much faster if the nucleophile is held in the right position for reaction. The hydrolysis of ethyl acid phthalate (6.4) is independent of pH in the range 5 to 7. The carboxylate group probably again acts as a nucleophile. The fact that these rates are independent of the pH shows that there is no catalysis by H^+ or OH^-, as there is in the usual hydrolysis of an ester.

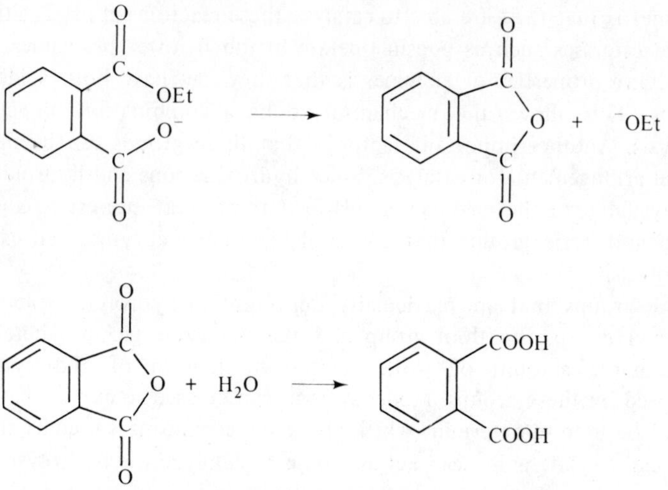

(6.4)

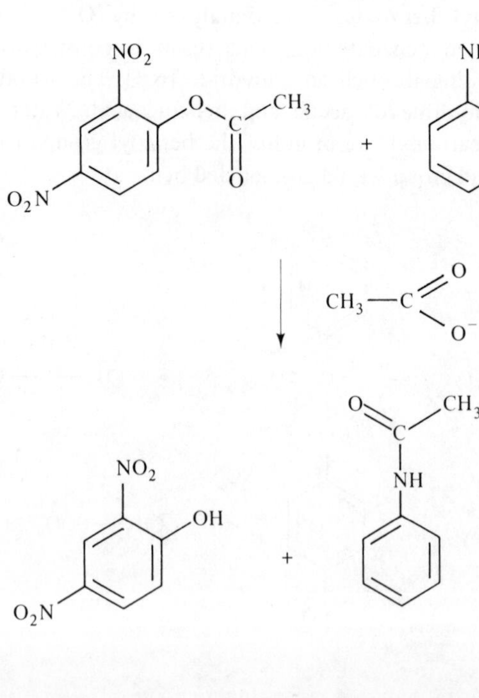

(6.5)

The anhydride intermediate in those reactions which proceed through anhydrides may be detected by the addition of aniline to the reaction mixture. When 2,4-dinitrophenylacetate (6.5) is hydrolyzed by acetate ion in the presence of aniline, the product is acetanilide. The source of the acetanilide must be from the acetic anhydride formed from the attack of the acetate ion on 2,4-dinitrophenyl acetate (6.6).

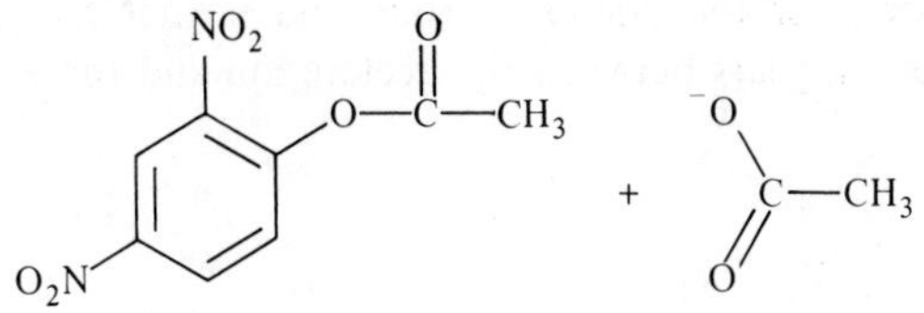

(6.6)

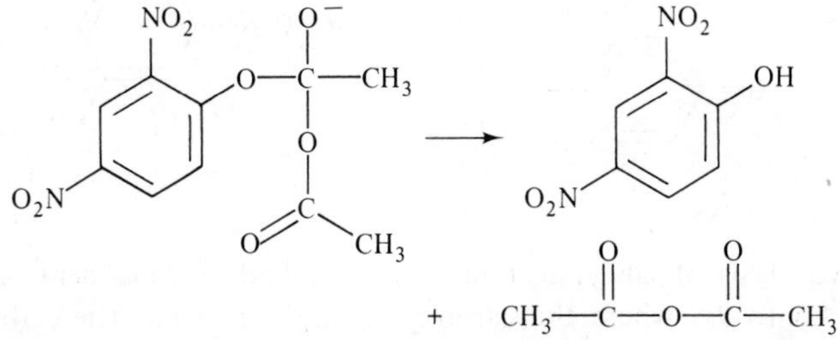

When phenyl acetate is hydrolyzed by acetate ion in the presence of aniline (6.7), no acetanilide is formed. With this ester, no acetic anhydride was formed: the acetate ion was acting in a different way. It is known that acetate ion acts as a general base to pull a proton off a water molecule, the resulting hydroxide ion attacking the ester as described for basic hydrolysis by hydroxide ion.

(6.7)

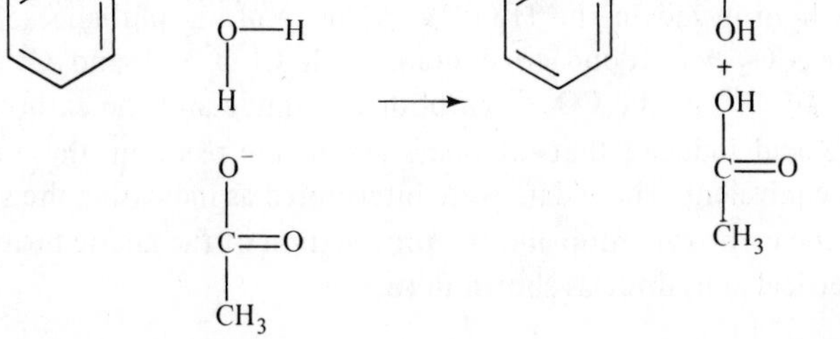

The question remains: why does acetate ion participate as a nucleophilic catalyst with one ester and a general base catalyst with the other? Acetate ion would add to each ester (6.8). Differences in mechanism are associated with the fate of the tetrahedral intermediate. This can collapse by the loss of a phenolate ion to give acetic anhydride or by the loss of acetate ion to reform the ester. Because the 2,4-dinitrophenolate ion is a weaker base than phenoxide ion, it is a better leaving group. Nucleophilic catalysis occurs. In intermediate (6.8b), the acetate ion leaves instead of the phenoxy group and no net reaction occurs. When a water molecule appears between the acetate ion and the ester, general base catalysis can occur.

$$\mathrm{Ar{-}O{-}C(O^-)(OAc){-}CH_3} \qquad (6.8)$$

(a) Ar = phenyl

(b) Ar = 2,4-dinitrophenyl (O_2N–, NO_2)

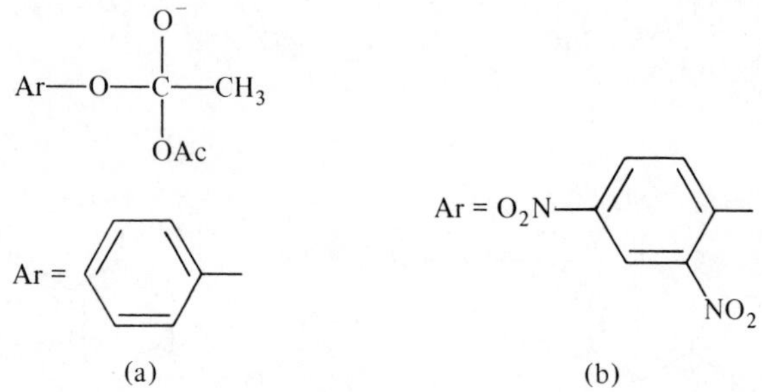

The hydrolysis of salicyl acetate (6.9) is similarly independent of pH in the range of 4.5 to 9.5 where the carboxyl group is ionized. The carboxyl group could act as a nucleophilic catalyst here also, but for reasons that we do not have space to discuss the belief is that the carboxylate ion acts as a general base and merely removes a proton from water to allow the water to attack. Thus a carboxyl group can act as a catalyst for ester hydrolysis in two different manners. It is not known which of these mechanisms occur in the enzymatic catalysis when a carboxyl group is involved.

Other esterase models have demonstrated the importance of holding the acidic group in the correct position. Phthalamic acid (6.10a) is hydrolyzed 10^5 times as fast as benzamide (6.10b) at pH 3. The rate is independent of the pH in the range of 1.3 to 2.6 but decreases at higher pH values. The undissociated phthalamic acid is therefore the reactive species, and hydronium ion is not involved in the mechanism. When phthalamic acid labeled with C^{13} in the amide group was hydrolyzed in the H_2O^{18} and the resulting phthalic acid decarboxylated, the CO_2 was found to contain both $C^{13}O^{16}O^{18}$ and $C^{12}O^{16}O^{18}$. The occurrence of O^{18} in the CO_2 from both the amide and the carboxyl group of phthalamic acid indicate that, at one stage in the reaction, these groups must have been equivalent. These data were interpreted as indicating the simultaneous attack by the carboxyl group and the protonation of the amide nitrogen to form the symmetrical anhydride as shown in (6.11).

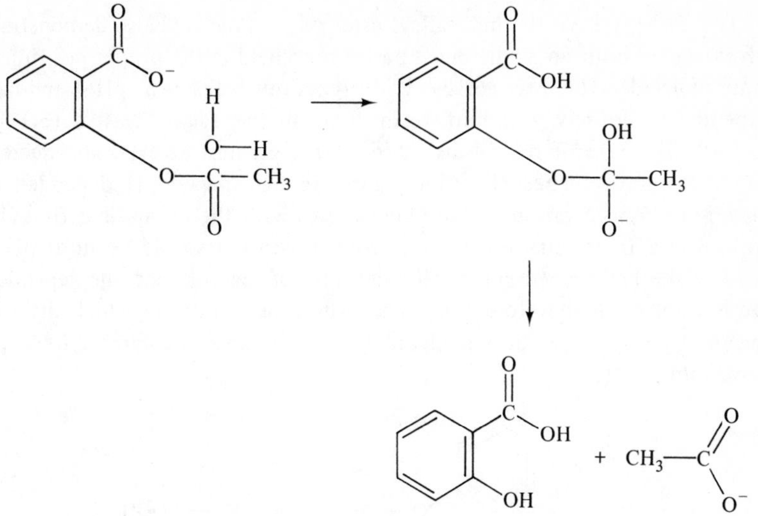

(6.9)

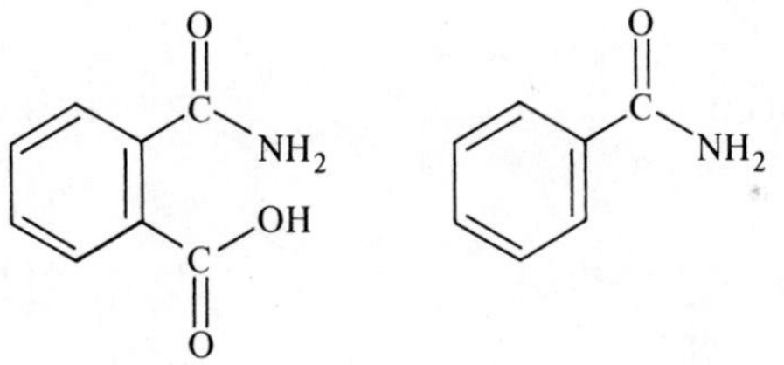

(6.10)

(a) (b)

(6.11)

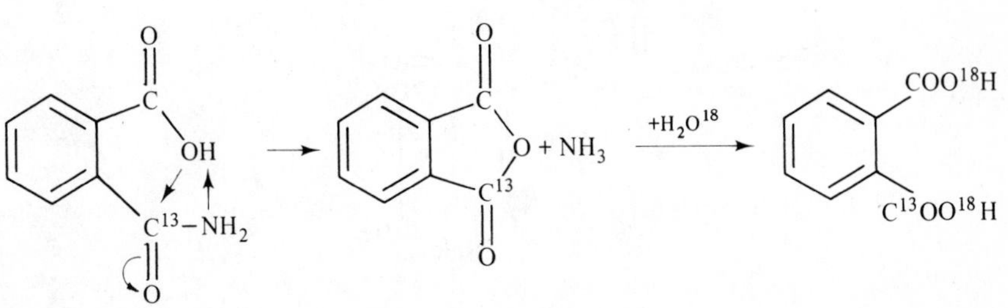

$+H_2O^{18}$

C^{13}–NH_2

$O + NH_3$

$COO^{18}H$

$C^{13}OO^{18}H$

The hydrolysis of the half-salicyl ester of succinic acid also demonstrates the advantage of both an acidic and a basic group held in the proper position on the same molecule. The ester (6.12a) hydrolyzes much faster at pH 4 and has a pH dependence entirely different from that of the related esters (6.12b) and (6.12c). The rate of hydrolysis of (6.12a) is greatest at pH 4 and decreases at either lower or higher pH values, giving a bell-shaped pH dependence. The maximum concentration of the ionic species with the aromatic carboxyl group ionized and the aliphatic carboxyl group unionized would occur at pH 4–the peak of the bell-shaped curve. The high rate of reaction and the dependence of the rate on pH is therefore consistent with a mechanism in which the aromatic carboxyl group serves as a nucleophile and the aliphatic carboxyl group as an electrophile (6.13).

$$\text{(a)}\quad o\text{-}(COOH)C_6H_4\text{—O—C(=O)—}CH_2\text{—}CH_2\text{—}COOH$$

(6.12)

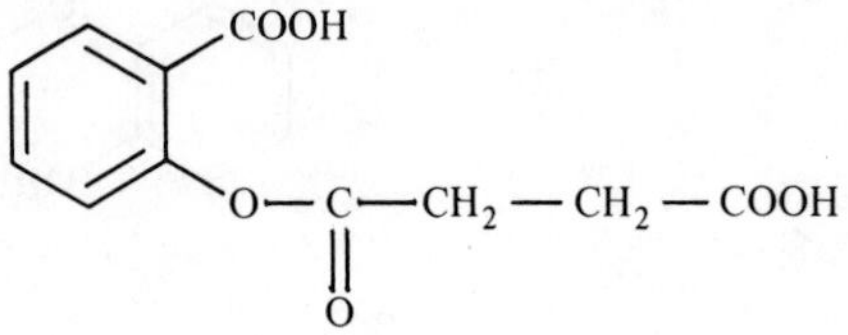

$$\text{(b)}\quad o\text{-}(COOH)C_6H_4\text{—O—C(=O)—}CH_3 \qquad \text{(c)}\quad o\text{-}(COOCH_3)C_6H_4\text{—O—C(=O)—}CH_2\text{—}CH_2\text{—}COOH$$

(6.13)

$$o\text{-}(C(=O)O^-)C_6H_4\text{—O}\cdots\text{H—O—C(=O)—}CH_2\text{—}CH_2\text{—C=O}\quad (O^- \rightarrow C)$$

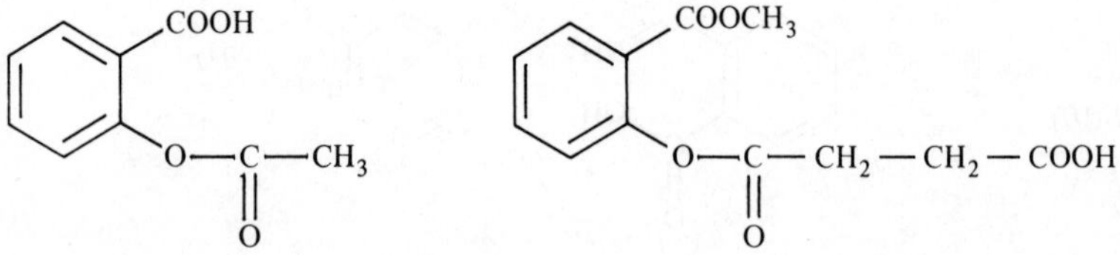

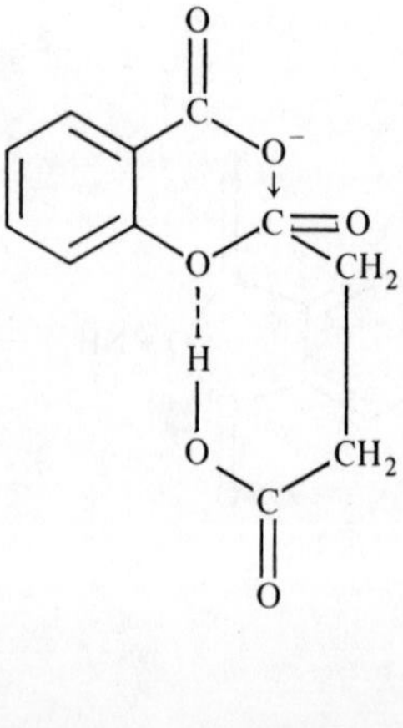

Imidazole will also act as a nucleophilic catalyst. This is interesting because histidine occurs in the active center of many esterases. Imidazole is a catalyst for the hydrolysis of phenylacetates. The imidazole acts as a nucleophilic catalyst in (6.14) in which the imidazole displaces the phenylate ion. The resulting acetyl imidazole is quite unstable and hydrolyses quickly in water.

(6.14)

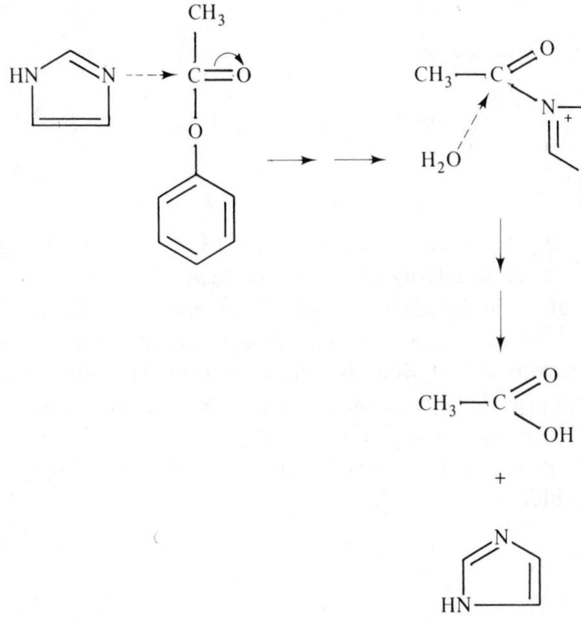

Thiol groups occur in proteins and have also been found to serve as nucleophilic catalysts. Ficin and papain are two proteinases that have thiol groups that are essential for their activity. Mercaptans will catalyze the hydrolysis of certain esters (6.15). In (6.16), the catalytically active species has been found by pH studies to be the dianion. In a proposed mechanism, the thiophenolate ion attacks the *p*-nitrophenyl acetate to form a thiol ester. The carboxylate ion catalyzes the hydrolysis of the resulting thiol ester by either general base or nucleophilic catalysis as described previously. The rate determining step is the hydrolysis of the thiol ester. A carboxylate ion is required for this step so that the p*K* dependence of the overall reaction rate shows an inflection at approximately pH 4. Interestingly, papain and ficin also show inflections in their rate versus pH curves at approximately pH 4. Various

(6.15)

(6.16)

$$\text{C}_6\text{H}_4(\text{S}^-)(\text{COO}^-) + O_2N\text{-}C_6H_4\text{-}O\text{-}\overset{O}{\overset{\|}{C}}\text{-}CH_3 \longrightarrow \text{C}_6\text{H}_4(\text{S-}\overset{O}{\overset{\|}{C}}\text{-}CH_3)(\text{C(=O)-O}^-)$$

mechanisms utilizing the nucleophilic properties of sulfur have been suggested to account for the enzymatic activity of papain and ficin.

Nucleophiles also catalyze the hydrolysis of phosphate esters. An anion formed close to, or held close to, the phosphorus may have a very great advantage over anions in solution that have to overcome the negative shell around the phosphorus. Salicyl phosphate, which has a carboxyl ion held close to the phosphorus, is hydrolyzed quite rapidly through (6.17). The attacking group is the carboxylate ion and the reaction proceeds through benzoyl phosphate anhydride.

(6.17)

$$\text{C}_6\text{H}_4(\text{O-P}^+(\text{O}^-)_2\text{-OH})(\text{C(=O)O}^-) \longrightarrow \text{C}_6\text{H}_4(\text{O}^-)(\text{C(=O)-O-P}^+(\text{O}^-)_2\text{-OH}) \longrightarrow \text{Salicylic acid} + \text{Phosphate ion}$$

6.3 ELIMINATION REACTIONS

Elimination reactions are also catalyzed by the group of enzymes without any cofactors. These reactions are essentially acid-base type reactions. We will discuss the various types of elimination of HX from CH–CX to form an olefin. The

(6.18)

$$>CH\text{—}CX< \xrightarrow{-X^-} >CH\text{—}C^+< \xrightarrow{-H^+} >C{=}C<$$

anion, X^-, may be lost first to produce a carbonium ion by (6.18); or the proton may be lost first to form a carbanion by (6.19); or both may be lost simultaneously as in (6.20). For the carbonium ion mechanism or the carbanion mechanism, either step may be rate determining so that a total of five different mechanisms are possible.

$$\text{(6.19)} \qquad \gt CH-CX \lt \xrightarrow{-H^+} \gt C^- -XC \lt \xrightarrow{-X^-} \gt C=C \lt$$

$$\text{(6.20)} \qquad \gt CH-CX \lt \xrightarrow{-H^+,\ -X^-} \gt C=C \lt + HX$$

It is possible to distinguish between at least four of these mechanisms by means of deuterium tracers. Two measurements can be made. The rate of exchange of CH–CX with D_2O can be compared with the rate of the overall reaction to olefin. If exchange is faster than olefin formation, this indicates that an intermediate ion is formed. This type of measurement must be interpreted with caution in enzymatic systems. There are examples in which the substrate rapidly exchanges deuterium with the enzyme, but little or no deuterium exchange between substrate and solvent is observed. If a proton donated to the enzyme is only slowly exchanged with solvent deuterium, that proton may be placed back on the product. Thus the lack of deuterium exchange between substrate and solvent is not always indicative that carbanions are not involved in the mechanism. Consider the hypothetical rearrangement (6.21) which proceeds through a carbanion. If the enzyme donates back to the substrate the same proton that was removed, the major product will not contain deuterium even though the reaction is carried out in D_2O solvent.

The other measurement is the deuterium rate effect, i.e., does CD–CX eliminate DX more slowly than, or at the same rate as, CH–CX eliminates HX. A deuterium rate effect occurs when the C–H bond is broken in the slow, rate-determining step. Although carbanion, carbonium ion, and concerted reactions are all found in nonenzymatic reactions, nature seems to prefer carbanion elimination reactions. As we discussed in Chapter 3, all of the pyridoxal catalyzed eliminations have intermediates that are essentially carbanions in which the carbanion is stabilized by the positive quaternary nitrogen in pyridoxal.

When one considers the mechanisms of biological reactions in general, one is struck with the preponderance of carbanion reactions. One possible explanation of nature's preference for carbanion chemistry is that life originated under basic conditions. With an excess of ammonia in the primal soup, carbanion mechanisms were preferred over carbonium ion mechanisms.

$$
\begin{array}{ccc}
-CH-C(X)- + \text{Enzyme} & & \\
\updownarrow & & \\
-\overset{\ominus}{C}-C(X)- + H\text{—Enzyme} & \xrightarrow{D_2O} & -\overset{\ominus}{C}-C(X)- + D\text{—Enzyme} \\
\updownarrow & & \updownarrow \\
>C{=}C< + X^- + H\text{—Enzyme} & & >C{=}C< + X^- + D\text{—Enzyme} \\
\updownarrow & & \updownarrow \\
-C(X)-\overset{\ominus}{C}- + H\text{—Enzyme} & & -C(X)-\overset{\ominus}{C}- + D\text{—Enzyme} \\
\updownarrow & & \updownarrow \\
-C(X)-CH- + \text{Enzyme} & & -C(X)-C(D)- + \text{Enzyme} \\
\text{Major product} & & \text{Minor product}
\end{array}
\qquad (6.21)
$$

There are a few examples of elimination reactions via carbonium ion intermediates in nature. Examples are the fumarase reaction, which dehydrates malic acid to form fumaric acid, and the aconitase reaction. There appear to be no concerted elimination reactions in nature.

Carbonium ion mechanisms occur when the group X leaves first. The carbonium ion is formed in a rate-determining step and decomposes more rapidly by the loss of a proton to form the olefin. The carbonium ion is planar, and rotation of the single bond between the carbon atoms allows the most stable olefin isomer or mixture of isomers to be formed. An example of this reaction is the acid-catalyzed formation of an olefin from some alcohols. First the conjugate acid of the alcohol is formed and then this species loses water to form the carbonium ion (6.22).

The elimination of water from malic acid to form fumaric acid (6.23) is catalyzed by the enzyme fumarase. This reaction has been thoroughly studied both kinetically and mechanistically. The elimination catalyzed by the enzyme

$$ROH + H_3O^+ \rightleftharpoons RO^+H_2 + H_2O$$

(6.22)

$$\gt CH-C\lt-O^+H_2 \rightleftharpoons \gt CH-\overset{+}{C}\lt + H_2O$$

(6.23)

$$HOOC-HCOH-CH_2-COOH \rightleftharpoons HOOC-CH{=}HC-COOH + H_2O$$

was found to be *trans* by a study of the reverse reaction, the addition of deuterium oxide to fumaric acid. The *trans* position of the deuterium in the deuterated malic acid relative to the hydroxyl group was determined by nuclear magnetic resonance. Because deutero-1-malate loses HOD at close to the same rate as the nondeuterated compound loses H_2O, the hydrogen-carbon bond does not break in the rate-determining step. The rate-determining step must therefore be cleavage of the carbon-oxygen bond. A prior, fast removal of hydrogen ion has been shown not to occur because malate ion and fumerase do not incorporate deuterium into the malate ion faster than can be accounted for by the back reaction from fumaric acid. These observations are consistent with a rate-determining formation of a carbonium ion followed by a rapid loss of a proton. The retention of configuration found in the fumarase reaction at first seems in contradiction to a carbonium ion reaction. However, this is a common observation in enzymatic reactions. The carbonium ion is so highly hindered on an enzyme surface that there is a preferential side of reaction.

This reaction mechanism requires that the enzyme has two functional groups, an acidic group to add a proton to the hydroxyl and a basic group to pull the proton off carbon. The pH dependence of the fumerase reaction indicates that the enzyme does have two essential functional groups. The pK's of these groups were determined by a thorough study of the pH dependence of the fumarase reaction. The pK's were found to be 6.2 on the free enzyme, 5.3 and 7.3 in the enzyme-fumarate complex, and 6.6 and 8.4 in the enzyme-malate complex. These groups are possibly a carboxyl group and an imidazole group.

A similar mechanism occurs in the aconitase reaction. The aconitase-catalyzed labeling of citric acid in D_2O occurs at the same rate as the formation of aconitic acid. This result is in agreement with a carbonium ion mechanism similar to the fumarase reaction. This reaction has also been shown to be a *trans* elimination.

Carbanion mechanisms commonly occur when the carbanion is stabilized by conjugation with a carbonyl group of an aldehyde, ketone, ester, or acid.

Addition reactions, the reverse of the elimination reactions with this type of conjugated anions, are called *Michael reactions.*

The addition of ammonia to mesaconic acid, catalyzed by the enzyme β-methylaspartase, has been shown to proceed through an anion because the enzyme catalyzes a fast deuterium exchange between water and the product, β-methylaspartic acid. The reaction must, therefore, be a Michael reaction (6.24) in which ammonia is added to a conjugated acid. The reaction occurs at pH 8 so that the intermediate must have two negative charges on one carboxyl group. A dianion such as this would normally be quite hard to form, but the enzyme requires both K^+ ion and Mg^{++} ion. These metal ions presumably help to stabilize the dianion. A nuclear magnetic resonance study of the product of the reaction, β-methylaspartic acid, has shown that the elimination is *trans.*

$$(^-OOC)(H)C{=}C(CH_3)(C{=}O)O^- + NH_3 \rightleftarrows (^-OOC)(H)(NH_3^+)C{-}C(CH_3){=}C(O^-)O^- \quad (6.24)$$

$$\downarrow\uparrow + H^+$$

$$-OOC(H)(NH_3^+)C{-}C(H)(CH_3)COO^-$$

Crotonase catalyzes the elimination of water from β-hydroxybutyric acid to form crotonic acid. Crotonase will isomerize crotonic to isocrotonic acid. Crotonase contains essential sulfhydryl groups. The most likely mechanism for the overall reaction from crotonic acid to β-hydroxybutyric acid is an addition of the enzyme to the crotonic acid, presumably via a Michael reaction (6.25), followed by a displacement of the enzyme by water. The addition of the thiol group to the crotonic acid would proceed through an intermediate carbanion which would be stabilized by the thiol ester of coenzyme A, as described in Chapter 4.

Concerted 1,4-elimination reactions lose the two groups on the same side of the molecule in a *cis* manner. Prephenic acid undergoes two types of elimination reactions [(6.26) and (6.27)] to produce either *p*-hydroxyphenyl pyruvic acid or phenyl pyruvic acid. The hydroxyl group is *cis* to the carboxyl group allowing the elimination of water to be concerted. The trans elimination of carbon dioxide and a hydride ion cannot be concerted and need not be, because an

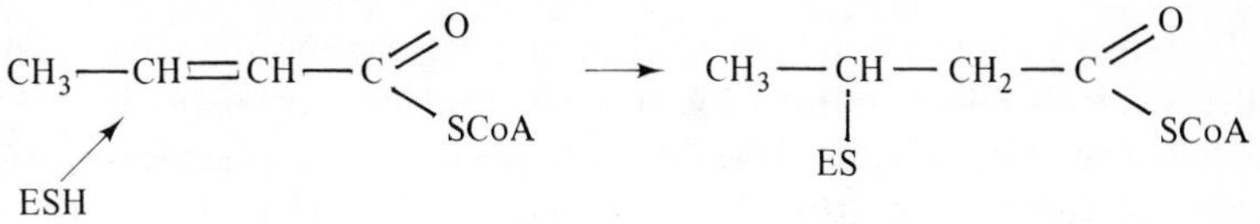

$$CH_3-CH=CH-C(=O)SCoA \xrightarrow{ESH} CH_3-CH(ES)-CH_2-C(=O)SCoA$$

(6.25)

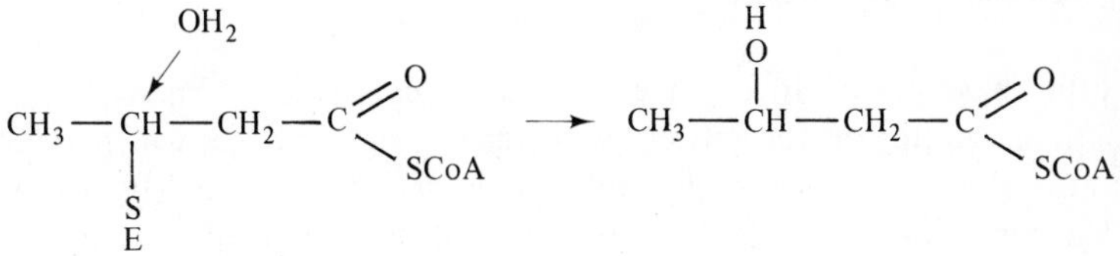

$$CH_3-CH(SE)-CH_2-C(=O)SCoA \xrightarrow{OH_2} CH_3-CH(OH)-CH_2-C(=O)SCoA$$

(6.26)

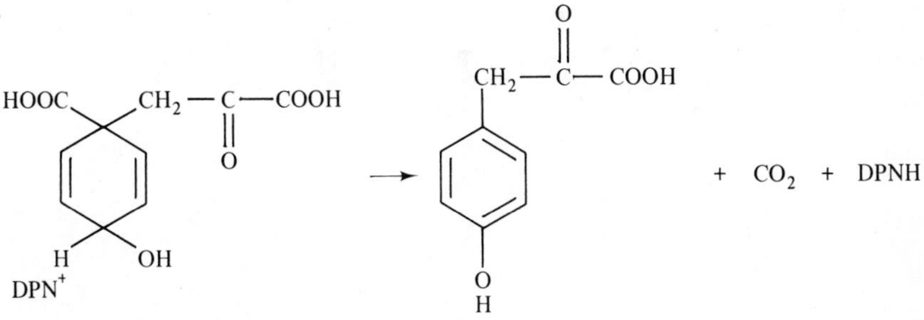

$$\text{HOOC}(\text{CH}_2-\text{CO}-\text{COOH})\text{-cyclohexadiene}(\text{H})(\text{OH}) \xrightarrow{DPN^+} p\text{-HO-C}_6\text{H}_4-CH_2-CO-COOH + CO_2 + DPNH$$

(6.27)

$$\text{HOOC}(\text{CH}_2-\text{CO}-\text{COOH})\text{-cyclohexadiene}(\text{H})(\text{OH}) \longrightarrow C_6H_5-CH_2-CO-COOH + CO_2 + H_2O$$

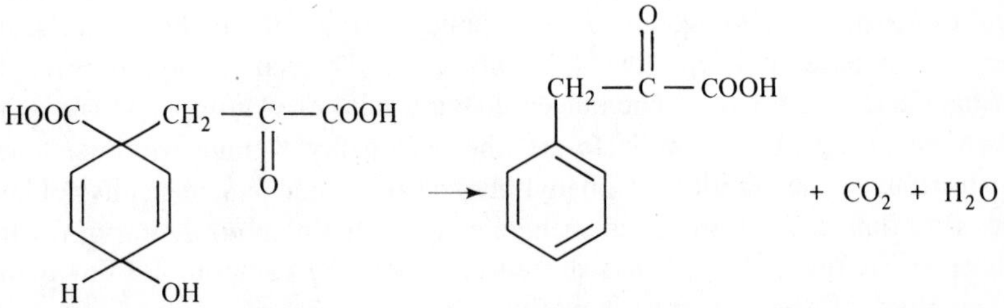

intermediate ketone can be formed which could decarboxylate in a second step. This again demonstrates the beauty of nature. The reaction that cannot proceed through an intermediate and that must be concerted has the leaving groups arranged in a *cis* manner. In the other reaction, the geometry is not mechanistically important. By elimination of the *cis* placement of substituents, the leaving groups are arranged *trans*.

That two important compounds can be formed from the same intermediate under stereochemical controls by enzymes provides an important economy to the cell. The pathways to tyrosine and to phenylalanine share several steps. The number of enzymes and the necessary genetic information to make the enzymes for forming these amino acids are minimized.

6.4 REARRANGEMENTS

Another example of an enzyme that catalyzes a reaction merely by means of functional groups on the protein is the enzyme glyoxylase, which catalyzes the rearrangement of glyoxal to glycollic acid in neutral solution. The rearrangement will occur in basic solution without the aid of an enzyme. The nonenzymatic reaction in D_2O produces glycollic acid with no nonexchangeable deuterium.

$$\text{H(O=)C–C(=O)H} \xrightarrow{{}^{-}\text{OD}} \text{H(O=)C–CH(OD)(O}^{-}\text{)} \longrightarrow \text{H–CH(O}^{-}\text{)–C(=O)OD} \longrightarrow \text{DO–CH}_2\text{–C(=O)O}^{-} \qquad (6.28)$$

This result is consistent with the hydride ion rearrangement (6.28). The reaction requires essentially two agents: a nucleophile to attack the carbonyl in the first step, and a base to keep the $-O^-$ ionized for the second step. Franzen has designed catalysts for this. The nucleophile is a sulfhydryl group, and the base is an amino group. These catalysts are the *N,N*-dialkyl-β-aminoethylmercaptans which catalyze the reaction of phenyl glyoxal to mandelic acid at physiological pHs. Reaction in D_2O produced mandelic acid containing no deuterium, which indicates that the hydrogen that shifted never left the compound. In methanol, the product of the reaction is methyl mandelate. These observations are all consistent with (6.29).

The *N,N*-diethyl-β-aminoethylmercaptan has a turnover number of 0.7 at 20° in methanol. This is increased to 6.7 in β-piperidylethylmercaptan (6.30). These turnover numbers are still quite insignificant compared to the turnover number of 35,000 for glyoxylase I. Glyoxylase I catalyzes the reaction between methyl glyoxal and glutathione to give the glutathione ester of lactic acid. Reaction in D_2O produces lactic acid containing no deuterium in agreement with the model system. Presumably the mercaptan group is furnished by the glutathione and the amino group by the enzyme. However, the enzyme must be doing something more than the model system because of the great discrepancy in the turnover numbers.

(6.29)

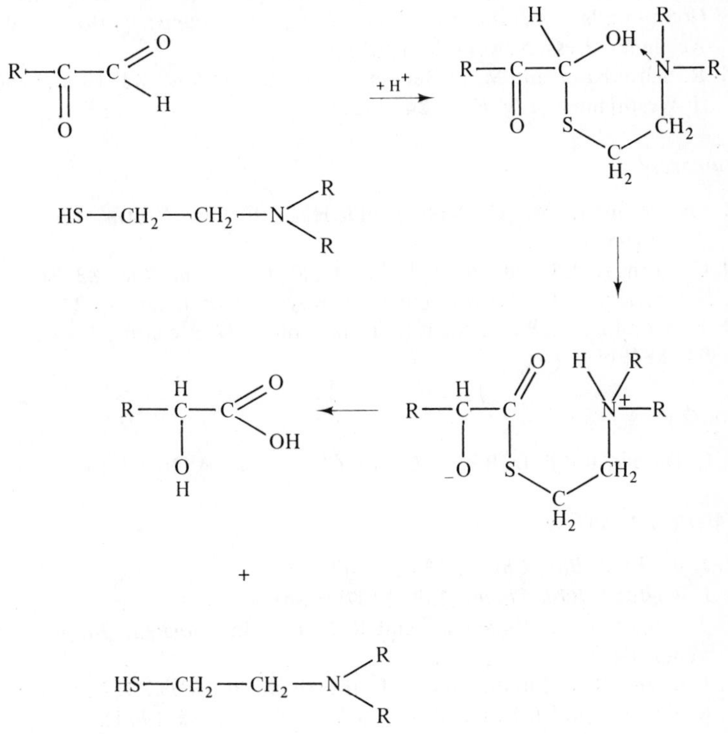

$$R-CO-CHO + HS-CH_2-CH_2-NR_2 \xrightarrow{+H^+} \text{cyclic adduct} \rightarrow \text{cyclic intermediate} \rightarrow R-CH(OH)-COOH + HS-CH_2-CH_2-NR_2$$

(6.30)

$$\text{(piperidino)}N-CH_2-CH_2-SH$$

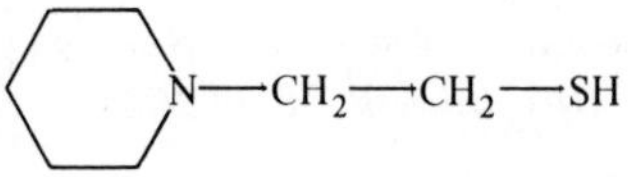

REFERENCES

Esterification and Hydrolysis

M. L. Bender, *J. Am. Chem. Soc.*, **79**, 1258(1957).
M. L. Bender and M. C. Neveu, *J. Am. Chem. Soc.*, **80**, 5388(1958).
T. C. Bruice, T. H. Fife, J. J. Bruno, and P. Benkovic, *J. Am. Chem. Soc.*, **84**, 3012(1962).

A. R. Ferscht and A. J. Kirby, *J. Am. Chem. Soc.,* **89**, 4853, 4857(1967).
A. R. Ferscht and A. J. Kirby, *J. Am. Chem. Soc.,* **90**, 5818, 5833(1968).
H. Lindley, *Adv. Enz.,* **15**, 271(1954).
D. G. Oakenfull, T. Riley, and V. Gold, *Chem. Comm.,* **12**, 385(1966).
D. Samuel and B. L. Silver, "Oxygen Isotope Exchange Reactions of Organic Compounds", *Advances in Physical Organic Chemistry,* Vol. 3, ed. V. Gold, Academic Press, New York (1965) p. 123.
G. R. Schonbaum and M. L. Bender, *J. Am. Chem. Soc.,* **82**, 1900(1960).
F. H. Westheimer, *Adv. Enz.,* **24**, 441(1962).

Fumarase

R. A. Alberty, W. G. Miller, and H. F. Fisher, *J. Am. Chem. Soc.,* **79**, 3973(1957).
O. Gawron, A. J. Glaid, and T. P. Fondy, *J. Am. Chem. Soc.,* **83**, 3634(1961).
J. N. Hansen, E. C. Dinovo, and P. D. Boyer, *J. Biol. Chem.,* **244**, 6270(1969).
D. E. Schmidt, Jr., W. G. Nigh, C. Tanzer and J. H. Richards, *J. Am. Chem. Soc.,* **91**, 5849(1969).

Enolase

E. C. Dinovo and P. D. Boyer, *J. Biol. Chem.,* **246**, 4586(1971).

β-Methyl Aspartase

H. J. Bright, *J. Biol. Chem.,* **239**,2307(1964).
H. J. Bright, *J. Biol. Chem.,* **239**, 2307(1964).
H. J. Bright, L. L. Ingraham, and R. E. Lundin, *Biochem. Biophys., Acta,* **81**, 576(1964).
H. J. Bright, R. E. Lundin, and L. L. Ingraham, *Biochem.,* **3**, 1224(1964).
V. R. Williams and J. Selbin, *J. Biol. Chem.,* **239**, 1635(1964).

1,4-Eliminations

R. K. Hill and G. R. Newkome, *J. Am. Chem. Soc.,* **91**, 5893(1969).
H. Plieninger, *Ang. Chem. (Int. Ed.),* **1**, 367(1962).

Glyoxylase models

V. Franzen, *Ber.,* **88**, 1361(1955).
V. Franzen, *Ber.,* **89**, 1020(1956).

7 METAL IONS IN BIOCHEMISTRY

7.1 INTRODUCTION

Certain metal ions are essential for the life of organisms. The elements N, S, O, P, C, and H are used in the building blocks of biological compounds, for amino acids, sugars, nucleic acids, and coenzymes. Their function is in organic compounds. In contrast, the metal ions are used primarily for maintaining charge neutrality, but, at times, they do possess catalytic functions. These include K^+ Na^+, Ca^{++}, and Mg^{++}. Potassium ions and magnesium ions are used for neutrality inside the cell, and sodium ions and calcium ions, for neutrality outside of the cell. There are certain trace elements that are used by the cell in catalysts. These include Mo, Zn, Cu, Mn, Fe, Co, V, and I. In this chapter we will discuss a few of these elements. The functions, or at least a function, of each of these elements are believed to be known at this time. There are other elements that may or may not be necessary for life. Certainly their function is not known at this time. These are F, Si, Cr, Se, and Sn. This chapter will essentially discuss the importance of the trace elements to life.

The transition from organic chemistry to inorganic chemistry is blurred in some areas. For instance, compounds that contain carbon-to-metal covalent bonds such as alkyl magnesium halides (Grignard reagents) are generally classified as organometallic compounds. In contrast, complex ions with metal ions bonded to organic ligands through heteroatoms, such as the EDTA (ethylenediamine tetraacetate) complex of lead (II), traditionally fall in the province of inorganic chemistry. (The II in parentheses after "lead" refers to its plus two oxidation state.) By these standards, this chapter contains biological examples of both organometallic and inorganic chemistry.

7.2 METAL IONS AS LEWIS ACIDS

There are many enzymes that contain no cofactor other than a metal ion. These are in addition to a number of metalloenzymes that contain organometallic cofactors such as iron porphyrins or cobalt corrins. Metal ions as cofactors commonly serve as general acids by virtue of their positive charge. Many reactions need an acid catalyst, and a metal ion serves as a convenient source of a general acid under physiological conditions. Metal ions are not as good acids as a proton because of their larger size. However, this is partially compensated for by a greater number of positive charges on the species. The metals usually used for acid catalysis by enzymes are magnesium, manganese, and zinc. Metals with partially filled *d* orbitals are usually not good acid catalysts because the electrons in the highly extended *d* orbitals tend to shield the positive charge. This problem is reduced in chelates if the ligands cause *d* orbitals away from the substrate to be filled and those toward the substrate to be empty.

A good example of a metal catalyzed reaction in which the metal acts as an acid catalyst is the enzymatic decarboxylation of oxalacetic acid. When an acid is decarboxylated, the carbon dioxide leaves without a pair of electrons (7.1). If the decarboxylation is to proceed at a practical rate, the pair of electrons left on the R group must be stabilized by some means. In a β-keto acid this is accomplished by placing the electrons on the carbonyl oxygen (7.2). At low pH values where the carboxyl group is unionized, the proton on the carboxyl group may aid the process by making the oxygen a better electron sink in a cyclic transition state (7.3). Divalent metal ions strongly catalyze the decarboxylation of α,α-dimethyloxalacetic acid. The metal ions act as an electron sink for the electron pair and thus act quite analogously to the proton. The first product of the decarboxylation is the metal chelate of the enol (7.4). The enol was observed as the first product by spectroscopy and by titration with bromine. The enzymatic decarboxylation of oxalacetic acid also requires divalent metal ions.

There are differences between the Mn^{++}-catalyzed and the enzyme-catalyzed decarboxylation of oxalacetic acid. The metal-catalyzed decarboxylation is about 6% slower if the carboxyl group contains C^{13}, whereas the enzymatic decarboxylation is the same for the C^{13} carboxyl labeled compound as for the

(7.1) $$R{-}CO_2^- \rightarrow R^- + CO_2$$

(7.2) $$R{-}C({=}O){-}CH_2{-}CO_2^- \longrightarrow R{-}C(O^-){=}CH_2 + CO_2$$

(7.3) $$R{-}C({=}O){-}CH_2{-}C({=}O){-}O{-}H \longrightarrow R{-}C(O{-}H){=}CH_2 + CO_2$$

(7.4) $$\text{(}^-O{-}C({=}O){-}C({=}O){-}C(CH_3)_2{-}CO_2^-\text{)}\cdots M^{++} \longrightarrow \text{(}^-O{-}C({=}O){-}C(O^-){=}C(CH_3)_2\text{)}\cdots M^{++}$$

C^{12}. The rate of the manganese catalyzed reaction is the same in D_2O as in H_2O, but the enzymic reaction is considerably slower in D_2O. Evidently carbon-carbon bond cleavage is rate determining for the reaction catalyzed by manganese ion but not for that catalyzed by the enzyme. The carbon-carbon bond breaks in the enzyme-catalyzed reaction more than 10^8 times as fast as in the reaction catalyzed by Mn^{++}.

The carboxylation reactions to form β-keto acids proceed by the reverse mechanism. The enol form is carboxylated to give the keto acid. That the enzymatic carboxylation of phosphoenol pyruvate first produces the keto and not the enol form was shown by carboxylating in deuterium oxide and showing that the keto acid contained no deuterium. An enol would have gained deuterium upon rearrangement to the keto acid.

Some hydrolytic enzymes also use metal ions as Lewis acids. An example is the enzyme carboxypeptidase, which catalyzes the hydrolysis of peptide bonds, and which contains zinc ion as an acid catalyst. The *in vitro* hydrolysis of amino acid esters and amides is catalyzed by metal ions; for example, copper, cobalt, and manganese all catalyze the hydrolysis of amino acid esters. The metal ion can act as a proton does by polarizing the carbonyl group and thus aiding attack by water (7.5). The hydrolysis of glycine amide is catalyzed by cupric ion.

(7.5) $$OH_2 \rightarrow R{-}CH(NH_2){-}C({=}O){-}OR \quad (H_2N\cdots M^{++}\cdots O)$$

The mechanism of action of carboxypeptidase is fairly well known by the X-ray analysis of the enzyme-substrate complex (7.6). The zinc ion acts as an acid catalyst, bonding to the carboxyl of the amide group on the substrate. The nitrogen of the amide is acidified by hydrogen bonding to a tyrosine side chain on the enzyme. The carboxyl group of a glutamic acid is about 2.5Å from the carbon of the carboxyl group. It is not possible to determine whether there is a water molecule between the carboxyl of the glutamic acid and the carboxyl of the peptide. The distance, 2.5Å, would accomodate a water molecule, but it would be a tight fit. Thus we do not know whether the carboxyl group is acting as a general base catalyst or, directly, as a nucleophile. If the water were present, it would be a general base catalysis.

(7.6)

R
CH_2
O
C — O^-
CH
Lys
Zn^{++}
Arg
NH
H
O
C
O
Glu
CH
Tyr
NH
R
Glu

Metals also act as Lewis acids in the hydrolysis of phosphate esters. The large number of negative charges on the oxygens around the phosphorus is an important factor in phosphate ester hydrolysis. These charges hinder the attack by a charged reagent (such as hydroxide ion) by charge-charge repulsion and of a nucleophile (such as water) by charge-dipole repulsion. Metals ions, for example magnesium and manganese, neutralize these charges and catalyze the reaction. Orthophosphate ion attacks adenosine triphosphate if the charges are neutralized by the manganous ion. The products formed are adenosine diphosphate and pyrophosphate ions. Almost all enzymes that catalyze reactions in which adenosine triphosphate or adenosine diphosphate are products or reactants contain magnesium ion as a cofactor. Undoubtedly, the magnesium ion acts as a general acid catalyst.

7.3 IRON PORPHYRINS

The structures of iron porphyrin compounds and of cobalt corrin compounds are given in (7.7). Iron porphyrins would be classed as bioinorganic compounds. A porphyrin ring without any side chains is shown in (7.7a). In contrast, cobalt corrins differ in both the type of macrocyclic ring structure and in the degree of saturation in these rings. There is commonly a base, B, on one side of the ring and an organic group, R, attached by a carbon-cobalt bond on the other side of the plane. The side chains on the corrin ring have been omitted in (7.7b). The carbon-cobalt bond makes coenzyme B_{12} an organometallic compound. The structural similarities of these cobalt corrins and of the iron porphyrins are great. The biological functions differ widely.

(7.7)

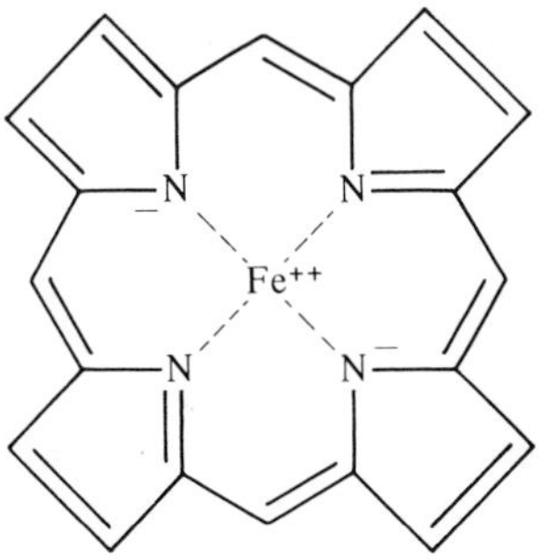

(a) 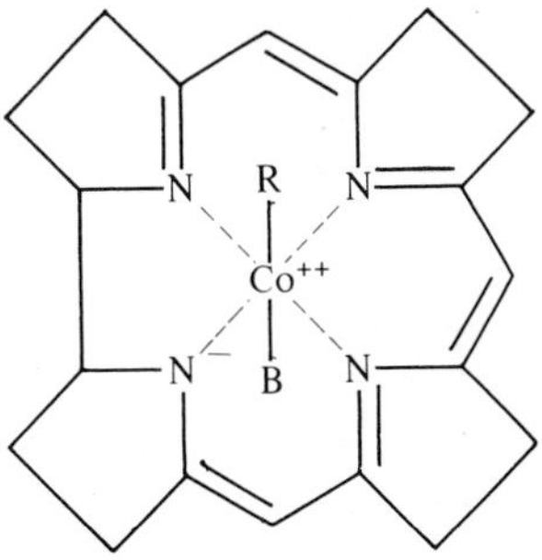 (b)

Iron porphyrins are the prosthetic group of hemoglobin and myoglobin—proteins that transport and store molecular oxygen; of the cytochromes—proteins that participate in the transport of electrons; and of catalase and peroxidase—proteins that remove toxic hydrogen peroxide from the body. Coenzyme B_{12} is a cobalt-corrin compound. Like the Grignard reagent in organic chemistry, it appears to be the source of powerful nucleophiles in biochemistry. These various functions are closely related to the electronic structure resulting from the interaction of the partially filled *d* subshell of the metal ion and the tetradentate porphyrin and corrin rings. The relationship of the electronic structure to the biological role will be explored in this chapter.

Crystal field theory, one approach to the modification of energy levels in transition metal complexes, will be used. (Elementary treatments can be found in many introductory general chemistry texts.) When six identical ligands are present, the electrons in *3d* orbitals are raised in energy to different extents.

Electrons in orbitals that place high electron density near the region occupied by electrons on the bases are raised higher than electrons in orbitals centered between ligands. The orbital splitting in an octahedral complex is given in (7.8).

(7.8)

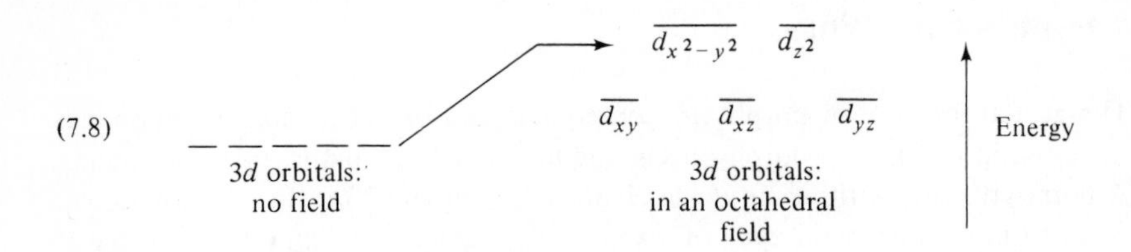

Biologically important complexes have less symmetry. Depending upon the particular ligands, an octahedral field may be distorted toward a square pyrimidal or square planar field. The corresponding energies are given in (7.9).

(7.9)

$d_{x^2-y^2}$
d_{z^2}
d_{xy}
d_{yz} d_{xz}

Square pyrimidal about *z* axis

Energy

$d_{x^2-y^2}$
d_{xy}
d_{z^2}
d_{yz} d_{xz}

Square planar about *z* axis

The porphyrin ring and the corrin ring bond to the metal ions using sp^2 hybridized orbitals on nitrogen. The nitrogen acts as a base donating an unshared pair to the Lewis acid. The remaining *p* orbital on nitrogen overlaps with *p* orbitals on carbons to form both bonding and antibonding orbitals having π symmetry. The bonding π orbitals are filled. Both the bonding and antibonding π orbitals have the proper symmetry to overlap with the d_{yz} and d_{xy} orbitals on the metal (7.10). The *d* orbitals are filled as are also the bonding π orbitals, so no interaction can occur between these orbitals. However, the filled *d* orbitals of the metal can donate electrons into the antibonding π orbitals of the ligand. The metal ion acts as a base, donating electrons back to the ligand. This back bonding delocalizes electrons in these *d* orbitals on the metal. It increases the strength of the total metal-ligand bonding by providing double bond character not available with sp^3 hybridized ligands such as aliphatic amines or water.

For entropic reasons, multidentate ligands bond more strongly than ligands having similar bond enthalpies. The ethylene diamine complex of copper (II) is dissociated to a far smaller extent than the corresponding complex with ammonia. The combination of the entropic advantage of the tetradentate ring and the enthalpic advantage associated with back bonding make the porphyrin and corrin rings hold very tightly to their central metal ions.

(7.10) $p-\pi^*$ d_{xy}

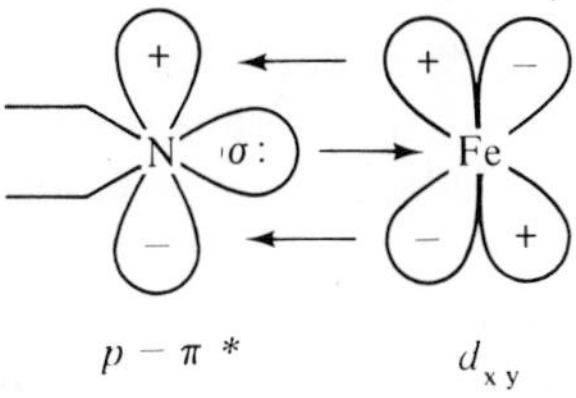

Hemoglobin contains iron in the plus two oxidation state. Ferrous ion has six electrons in the *3d* subshell. In an octahedral field, two electron configurations are possible. With strongly bonding ligands such as cyanide ion, ferrous ion forms a low-spin complex (7.11a). With weaker Lewis bases such as fluoride ion, the energy difference between orbitals is smaller, and a high-spin complex is favored (7.11b). Since a quantum of light excites an electron from a lower energy orbital to a partially or completely vacant upper level, the frequency of the visible light absorbed depends upon the energy difference between the subsets of *d* orbitals. The number of unpaired electrons is determined by magnetic measurements on the compound. Each unpaired electron makes a strong paramagnetic contribution to the net magnetism; paired electrons, a diamagnetic contribution. Spectroscopic and magnetic measurements are used to characterize a given complex.

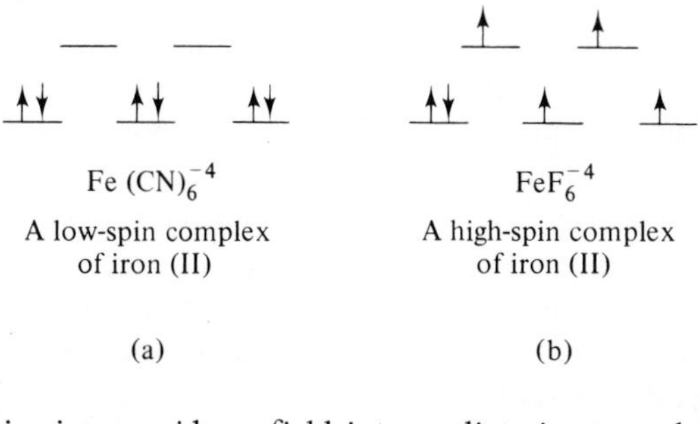

(7.11)

(a) $Fe(CN)_6^{-4}$ A low-spin complex of iron (II)

(b) FeF_6^{-4} A high-spin complex of iron (II)

The porphyrin ring provides a field intermediate in strength between that of high-spin ligands and low-spin ligands. (Yet because of entropy factors, iron is more tightly bonded to porphyrin than are stronger field, monodentate ligands.) Changes in the axial ligands, or of the bond distance to the histadine that forms an axial ligand in several heme proteins, apparently finely tune the field strength and resulting magnetic properties. Because these enzymes perform different functions, different properties are required. This chemical adaptability serves to make the iron porphyrins suitable for a large variety of enzyme functions.

Catalase and peroxidase catalyze the conversion of hydrogen peroxide to oxygen and water. The peroxide binds to iron at the active site of the enzyme, reacts, and the products are released. Rapid exchange of ligands is a desirable feature for these enzymes. In both catalase and peroxidase, the iron tends to be in a high spin state. This favors rapid exchange by either an S_N1 or an S_N2 mechanism.

Qualitative arguments for the contribution of spin type to the activation energy are given using iron (III) as an example. [This choice is arbitrary, Mossbauer spectroscopy indicates the presence of an iron (IV) state in both enzymes.] If the reaction is S_N2, the nucleophile must attack in a region between ligands. The electron density in this region is lower in the high-spin complex than in the low-spin species (7.12). The *d* orbitals concentrating electron density between ligands are doubly occupied in the low-spin state but only singly occupied in the high-spin structure. The extra electronic repulsion in the low-spin species would slow an exchange by this mechanism.

(7.12)

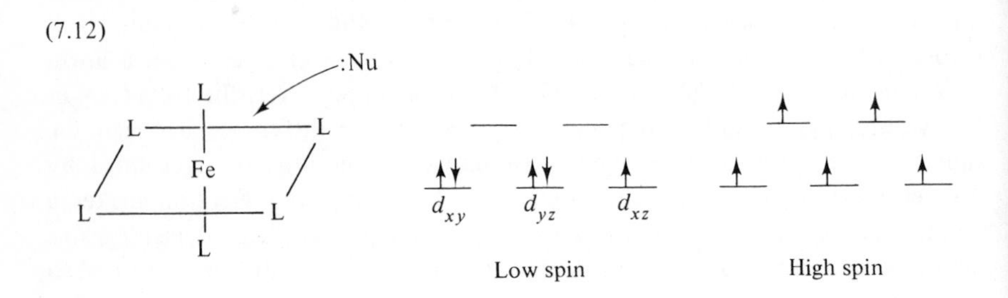

If the exchange occurs by an S_N1 mechanism, a change in geometry takes place in the rate determining step. An octahydral species is converted to a square pyramid as a group leaves. This change of geometry and the decreased number of ligands result in smaller splitting in the pentavalent species. The changes in the *d* orbital structures that would accompany this change are given in (7.13). In the low-spin species, electrons are raised in energy relative to the lowest energy orbitals. In the high-spin state there is a net decrease in energy. This difference of energy accompanying bond breaking and electronic reorganization appears as increased activation energy for the reaction of a low-spin species. The higher the activation energy, the slower the reaction.

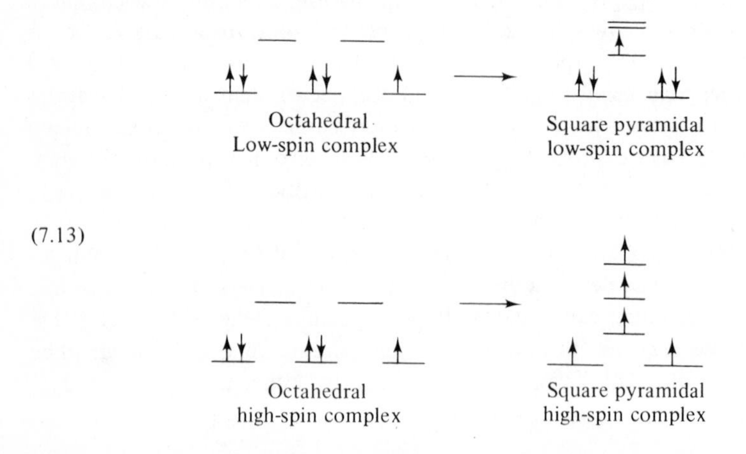

(7.13)

Catalase catalyzes the decomposition of hydrogen peroxide to oxygen and water. The kinetics show that catalase reacts first with one molecule of hydrogen peroxide to form an intermediate called complex *I*. This intermediate in turn reacts with another molecule of hydrogen peroxide to form the products, oxygen and water, and to reform the original enzyme. Catalase is in the ferric form before any hydrogen peroxide is added. A plausible structure for complex *I* is $Fe^{++}OOH$. Mossbauer spectra and ultraviolet spectra of complex *I* support iron +4. If $Fe^{++}OOH$ is the structure, there must be a large donation of an electron from the iron to the porphyrin ring. The ion $FeOOH^{++}$ is well known in inorganic chemistry from its charge transfer spectra in solutions containing ferric ion and hydrogen peroxide. This ion is comparable with the $FeOH^{++}$ ion.

There are many acid catalyzed oxidations by hydrogen peroxide. Two of these are the oxidation of organic sulfides to sulfoxides and amines to amine oxides. Essentially, the positive charge of the proton polarizes the oxygen-oxygen bond so that cleavage is ionic to water and, stoichiometrically, to OH^+. Iron is an acid catalyst and may act similarly. The reaction of complex *I* with another molecule of hydrogen peroxide to form water and oxygen may be another example of an acid catalyzed hydrogen peroxide oxidation (7.14). The group transferred to the OH^+ fragment would be a hydride ion. Hydride transfer in the rate-determining step is supported by a large rate difference when the reaction is run in D_2O and D_2O_2 solutions compared with the reaction in water.

$$\begin{array}{c} FeO{:}OH^{++} \\ H \\ \ddot{O} \\ \ddot{O} \\ \ddot{H} \end{array} \longrightarrow FeO^+ + H_2O + O_2 + H^+ \qquad (7.14)$$

The observation that both oxygen atoms in the molecular oxygen produced are derived from one hydrogen peroxide molecule and not from water, gives additional support for this mechanism. An alternate mechanism, involving FeO^{+++} as the iron species in complex *I*, provides another interpretation.

Peroxidase catalyzes the oxidation of various organic substrates by hydrogen peroxide. Peroxidase appears similar in overall reactions to catalase but differs drastically in mechanism. Peroxidase mechanisms proceed through free radical intermediates. Kinetic studies show that complex *I* is formed as in the reaction of catalase, but it now adds one electron at a time from substrates to form water and the original enzyme. The intermediate occurring after the addition of the one electron is called complex *II*. Complex *II* would have the oxidation level of ferryl ion, $Fe^{++}O$. Both complex *I* and complex *II* appear to contain +4 iron from spectral studies.

If we write $Fe^{++}OOH$ for complex *I* and FeO^{++} for complex *II*, we have (7.15). According to this mechanism, iron functions in an entirely different manner than in the catalase reaction. The iron now functions, by virtue of its

(7.15)
$$\begin{aligned} Fe^{+++} + H_2O_2 &\rightarrow Fe^{++} + OOH + H^+ \\ Fe^{++}OOH + SH_2 &\rightarrow FeO^{++} + SH\cdot + H_2O \\ FeO^{++} + H^+ + SH_2 &\rightarrow Fe^{+++} + SH\cdot + H_2O \\ 2SH\cdot &\rightarrow S + SH_2 \end{aligned}$$

(7.16)
$$H_2O_2 + e^- \rightarrow HO\cdot + {}^-OH$$

unpaired electrons, to stabilize the hydroxyl radical required for a one-electron reduction of hydrogen peroxide. One-electron oxidations by hydrogen peroxide produce unstable hydroxy radicals (7.16).

If the radical can be stabilized, the electron transfer will be more favorable. Ferric ion is paramagnetic and can stabilize a hydroxyl radical by pairing the extra electron of the hydroxyl radical with one of the iron. The one-electron oxidation by complex *I* produces ferryl ion (7.17), the stabilized form of a hydroxyl radical (7.18). The most interesting question remains as to why complex *I* acts as a two-electron oxidant in catalase and a one-electron oxidant in peroxidase? This question has not yet been answered, but the answer undoubtedly lies in the ligand field differences of the two enzymes.

(7.17)
$$Fe^{+++} + \cdot OH \rightarrow Fe^{++}\ OH^- \rightarrow FeO^{++} + H^+$$

(7.18)
$$H^+ + Fe^{++}OOH + e^- \rightarrow FeO^{++} + H_2O$$

In contrast to catalase and peroxidase, the cytochromes tend to be low spin species. Both out-of-plane ligands are contributed by the protein. The cytochromes participate in electron transport from the reduced nicotine adenine dinucleotides (NADH and NADPH) and reduced flavins ($FADH_2$ and $FMNH_2$) to oxygen. The ferrous-ferric redox couple is employed. The potential is modified by the ligands. The standard potential for different cytochromes in the electron transport chain differ by up to 0.20 V.

The rigid porphyrin ring is important for rapid electron transfer. Iron (III) with a smaller size and higher charge should be a stronger acid than iron (II). Ligands would bond more tightly to the ferric ion. The changes of bond distances following the electron transfer would be slow relative to the electron transfer. Electron transfer is instantaneous, hence atomic motions become rate determining. The rigid ring cannot change geometry. The potential slow step of bond reorganization is partially avoided by using the rigid, planar, tetradentate ring, because the only bond distance changes permitted are those to axial ligands. Both oxidized and reduced forms have the same iron to nitrogen distance in the plane of the ring.

Both hemoglobin, the oxygen transport protein, and myoglobin, the storage protein, contain iron (II) in a porphyrin ring. Molecular oxygen is transported as the sixth ligand on the heme ring. Oxygen diffusing from the hemoglobin in the red blood cells binds to myoglobin in muscle cells. Diving mammals such as the whale and seal are particularly rich in myoglobin.

If the iron of hemoglobin is oxidized to the ferric state, oxygen can no longer be carried. Nonetheless, the treatment of cyanide poisoning–inhalation of amyl nitrate–does oxidize hemoglobin. Cyanide is so very toxic because it inhibits an enzyme, cytochrome oxidase, involved in the production of ATP in oxidative phosphorylation. Iron (III) is a stronger acid toward cyanide than iron (II). Oxyhemoglobin binds the cyanide strongly. A blood transfusion can be used to complete the therapy.

In hemoglobin, the spin state is a function of the sixth ligand. With water present, hemoglobin is in the high-spin state. When a diamagnetic carbon monoxidide molecule or a paramagnetic oxygen molecule replaces water, a complete spin reorganization occurs to give a diamagnetic species. A closer look reveals that the change is still more complex. It has been shown, using X-ray crystallography, that the iron atom is 0.5 Å above the plane of the ring in the high-spin complex. In the low-spin complex, it lies in the plane of the ring.

The change of geometry upon binding oxygen is connected to the allosteric properties of hemoglobin. Hemoglobin consists of four associated polypeptide chains. Four heme rings provide binding sites for oxygen. These sites show cooperative behavior. If a site binds O_2, there is an additional probability that each additional site will bind oxygen. Changes in conformation are communicated from one chain to another, providing a mechanism for the heme sites to act cooperatively. This coordination of binding sites for oxygen is advantageous. When passing by the lungs, hemoglobin loads up efficiently. In the capillaries away from the lungs, it dumps oxygen efficiently.

7.4 FERRIDOXINS

In addition to the iron porphyrin compounds, there are a large number of nonheme proteins which act as oxidation-reduction catalysts. Little can be said at this time concerning their mechanism of action, but structurally they are of interest. They are also interesting because most of these proteins are strong reducing agents. They all contain relatively large amounts of sulfur and iron. They are also characterized by the evolution of H_2S in acidic solution. This is referred to as *acid labile sulfur.*

The simplest one is rubredoxin occurring in *Clostridium pasteurianum.* It contains one iron atom per molecule and no acid labile sulfur. The active center consists of an iron atom chelated by the mercapto groups of four cysteine amino acids in a tetrahedral arrangement as shown in (7.19). Rubredoxin is a one-electron oxidation-reduction catalyst with a potential −57 mV. The iron is high-spin ferrous in the reduced enzyme state and high-spin ferric in the oxidized form.

(7.19)

There are also nonheme iron proteins with two moles of iron atoms per mole of enzyme. These occur in all three kingdoms – bacteria, plants, and animals. The bacterial compound is called putidaredoxin, the plant compound is called chloroplast ferridoxin, and the animal form, adrenodoxin. These compounds contain two acid labile sulfurs in addition to the two iron atoms. The active center again has four cysteines. There is a tetrahedral arrangement of sulfurs about each iron (7.20). Again these compounds are one-electron redox catalysts. In the oxidized form, both iron ions are in a high-spin ferric state. In the reduced state, one iron is a high-spin ferrous ion and the other, a high-spin ferric ion. At low temperatures (below 100°K) there is coupling between the two iron atoms.

(7.20)

The chelate becomes low-spin in both the oxidized and reduced states with a spin of 1/2 in the reduced state and 0 in the oxidized state. Again, these enzymes are strong reducing agents with a potential of −430 mV at pH 7.

There are two types of enzymes (which have similar structures) with clusters of four iron atoms but which differ in their oxidation-reduction potentials. In addition to the four iron atoms in each cluster, there are four cysteine residues and four acid labile sulfurs (7.21). Both enzymes are one-electron redox catalysts. They appear to differ in their oxidation levels. The best guess is that the stronger oxidizing agent has one ferrous iron and three ferric irons in the oxidized state and two ferrous and two ferric irons in the reduced state. The stronger reducing agent has three ferrous ions and one ferric ion in the reduced state and two ferrous ions and two ferric ions in the oxidized state. The enzyme

(7.21)

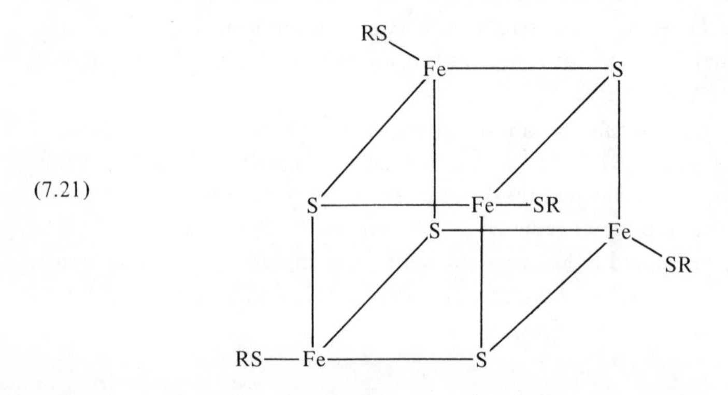

with a positive potential is called High Potential Iron Protein, or HPIP. It is found in photosynthetic bacteria. The redox potential is +350 mV at pH 7. The strong reducing agent of similar structure but probably of a different redox level is called bacterial ferridoxin because it is also found in bacteria. Its potential is highly reducing, −570 mV at pH 7. There are also eight iron ferridoxins found in clostridial bacteria. These are composed of two four-iron clusters. There are eight irons and eight acid labile sulfurs. The whole molecule is a two-electron oxidation-reduction catalyst with a potential of −420 mV at pH 7.

7.5 COENZYME B_{12}

At this point, a portion of the importance of particular ligands to structure and reactivity has been presented. Much of the area has barely been touched. Because the emphasis of this book is on bioorganic mechanisms, the authors choose to terminate this discussion and to consider coenzyme B_{12} in greater detail. This discussion is reinforced by the authors' research interests in B_{12}.

The discussion that follows is tentative. B_{12} chemistry remains controversial. It is the viewpoint of the authors that developing parallels to organic reaction mechanisms is an illuminating way to approach biochemical reactions. This approach is followed in this section.

Coenzyme B_{12} resembles iron porphyrin compounds. A transition metal ion is bound through nitrogen in a tetradentate ring with axial substituents above and below the ring. However, the metal ion is cobalt, not iron. The ring is a corrin rather than a porphyrin. Finally, and most dramatically, one ligand is bonded to the metal through carbon rather than through a heteroatom. While iron porphyrins are involved in a variety of enzymatic reactions, the known functions of cobalamines (B_{12}) involve powerful nucleophilic reactivity. The structural suitability of cobalamine for this function will be discussed by comparing its structure to iron porphyrins and to Grignard reagents.

The formation of a Grignard reagent from magnesium and an alkyl halide (7.22a) involves the conversion of carbon from an electron deficient state to an

$$CH_3Br + Mg \xrightarrow[\text{ether}]{} CH_3MgBr \quad \text{(a)}$$

$$CH_3MgBr + H_2O \longrightarrow CH_4 + Mg(OH)Br \quad \text{(b)}$$

(7.22)

$$CH_3MgBr + CO_2 \longrightarrow CH_3CO_2^- \; Mg^{++} \; Br^- \quad \text{(c)}$$

$$2CH_3MgBr + CdBr_2 \longrightarrow (CH_3)_2Cd + 2MgBr_2 \quad \text{(d)}$$

electron rich state. A change in oxidation number of carbon from -2 to -4 accompanies the reaction. The metal alkyl is so reactive that it is stable only in ether solvents. With acids such as water, the alkyl group is immediately protonated (7.22b). In a reaction of important synthetic utility, the Grignard reagent adds as a nucleophile to polar multiple bonds as is shown in (7.22c). Finally, the Grignard reagent reacts with salts of metals less reactive than magnesium to form metal alkyls that cannot be made directly (7.22d).

The reactions shown in (7.23) correspond closely to those given for the Grignard reagent. [B_{12}(s) refers to a reduced cobalamin. Two oxidation states of the coenzyme are known.] Reaction (7.23a) corresponds to the formation of the Grignard reagent. The methyl group is reduced by cobalt. In methyl tetrahydrofolic acid, it is bonded to nitrogen, an element more electronegative than carbon. In the organometallic compound, it is bonded to cobalt, an element less electronegative than carbon. Reaction (7.23b) accounts for the production of methane (swamp gas) in decaying vegetation, a reaction catalyzed by *Methanobacillus omelianskii* and by *Methanosarcina barkin. Clostridium thermoaceticum* fixes carbon dioxide by converting it to acetate (7.23c). Finally unicellular organisms are involved in mercury pollution. They methylate mercury (II) to form highly soluble methyl mercury compounds (7.23d). [They may also react with mercury (I) compounds.]

(7.23)

$$\text{Co } (B_{12s}) + N^5\text{—Methyltetrahydrofolic acid} \longrightarrow CH_3\text{—Co} + \text{Tetrahydrofolic acid} \quad \text{(a)}$$

Methyl B_{12}
(Methylcobalamine)

$$CH_3\text{—Co} + H_2O \longrightarrow CH_4 + \text{Co} \quad \text{(b)}$$

$$CH_3\text{—Co} + CO_2 \longrightarrow CH_3COOH + \text{Co} \quad \text{(c)}$$

$$CH_3\text{—Co} + HgCl_2 \longrightarrow CH_3HgCl + \text{Co} \quad \text{(d)}$$

Having looked at the correspondence of B_{12} to the Grignard reagent, it is useful to explore the structural suitability of cobalamines to that role. Since reactions of alkyl cobaloximes (7.24) resemble those of cobalamines and because the cobaloximes are frequently more easily studied, experimental evidence from this system will often be cited. Alkyl cobaloximes are prepared by the reaction sequence shown in (7.24).

$$Co^{++}_{(aq)} + 2\ CH_3C(=N{:}OH)C(=N{:}OH)CH_3 + 2OH^- \longrightarrow Co(II)\ (DMG)_2$$

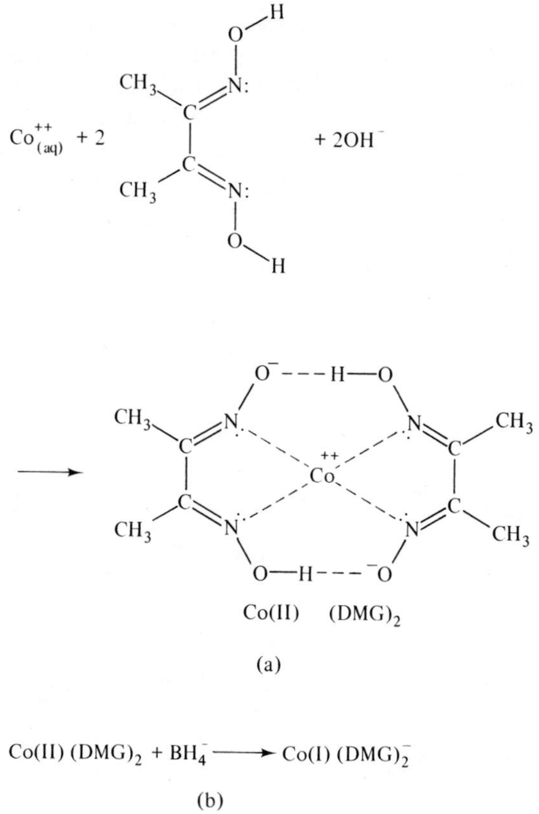

(a)

(7.24)

$$Co(II)\ (DMG)_2 + BH_4^- \longrightarrow Co(I)\ (DMG)_2^-$$

(b)

$$Co\ (I)\ (DMG)_2^- + RX \longrightarrow RCo\ (III)\ (DMG)_2 + X^-$$

(c)

The reactivity of cobalt (I) species in nucleophilic displacement reactions and the nature of the cobalt-carbon bond are both surprising. Reduced cobaloxime and cobalamine react with methyl iodide more than 10^{14} times as fast as methanol does. The resulting bond is long, 2.1 Å, compared to the carbon-carbon single bond distance of 1.54 Å. Even though the bond is long, the methyl group of methylcobalamine is electron rich. Its protons are more shielded,

($\tau = 10.1$) than those of *n*-propane ($\tau = 9.1$) when measured by nuclear magnetic resonance (NMR).

The Co (I) electronic structure in cobalamine or cobaloxime is not a simple d_8 configuration. Because of overlap with the π system of the ligands, the 4*p* orbital is lowered in energy relative to simple octahedral complexes. The partial use of the long 4*p* orbital (in a hybrid orbital) and the polarizability accompanying delocalization of electrons into the ring help account for the high reactivity. A striking example is the reaction of a reduced cobaloxime with a neopentyl bromide (7.25). In contrast, neopentyl bromide reacts very slowly with ethoxide ion (7.26). When silver ion is added, the rate increases, but the mechanism changes to S_N1. The speed of reaction of the Co (I) compounds is associated with high polarizability of long, delocalized orbitals. Bonding at long distances is associated with the less crowded transition state for the cobaloxime than for ethoxide, as shown by the rates of (7.25) and (7.26).

(7.25)

DMG Co (I) DMG + (cyclohexanone bearing $COOC_2H_5$ and CH_2Br; O)

fast →

(cyclohexanone bearing $COOC_2H_5$ and CH_2; O), CH_2 bonded to DMG Co DMG

(7.26)

$$(CH_3)_3CCH_2Br \xrightarrow[\text{very slow},\ S_N2]{^-OC_2H_5} (CH_3)_3CCH_2OC_2H_5$$

$$(CH_3)_3CCH_2Br \xrightarrow[S_N1]{Ag^+} (CH_3)_3CCH_2^+$$

Rearrangement and elimination ↓

$$(CH_3)_2C{=}CHCH_3$$

The use of a hybrid orbital with 4*p* character helps to account for the long bond distance in alkyl cobalamines and cobaloximes. The electronic distribution in cobalamines have been qualitatively described in a number of ways (7.27).

(7.27) R^+ Co(I) (a) R· Co(II) (b) R^- Co(III) (c)

Structure (7.27c) best accounts for (7.23b–d). However the strength and stability of the cobalt-carbon bond is greater than this structure implies. The Grignard reagent is too basic to exist in hydroxylic solvents; methylcobalamine is stable even in acid.

To be biologically useful, the nucleophilic reactivity of B_{12} needs to be triggered by specific enzymes. The mechanisms by which this occurs are not yet known with certainty. Experiments with model compounds and with enzymes are suggestive of mechanism, but are not yet conclusive.

Irreversible cleavage to give anions is possible using dithiols as reducing agents. (If the enzyme uses a redox reaction, it must be reversed in a later step if the enzyme is to function as a catalyst.) The production of methane and carbon dioxide from methyl cobaloxime was accomplished in this manner (7.28).

(7.28)

CH_3 DMG Co DMG Py + SH SH

→ [CH_3 DMG Co DMG S S H] → CH_3^- + DMG Co DMG + S S

CH_3^- —protic solvent→ CH_4

CH_3^- —CO_2 atmosphere, aprotic solvent→ CH_3COOH

(7.29)

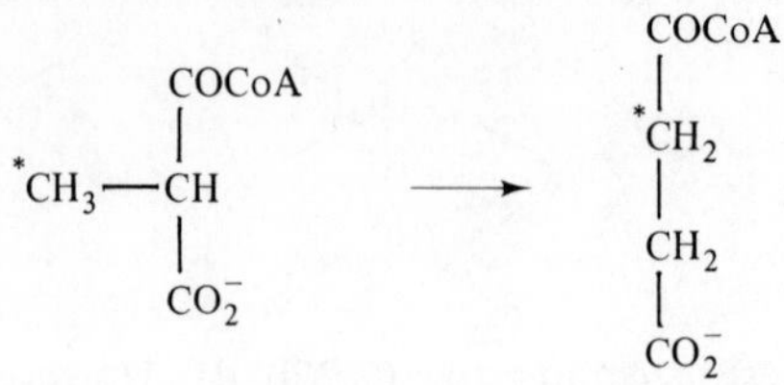

$$^{*}CH_3-CH(COCoA)(CO_2^-) \longrightarrow COCoA-{}^{*}CH_2-CH_2-CO_2^-$$

(7.30)

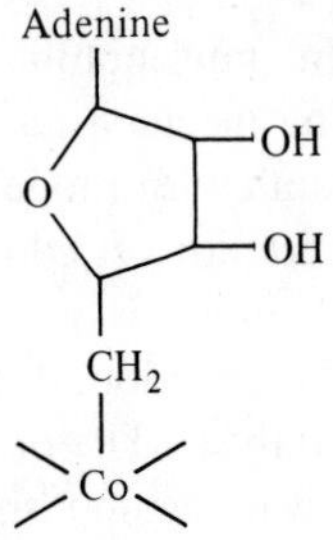

Adenine

OH

O

OH

CH_2

Co

(7.31)

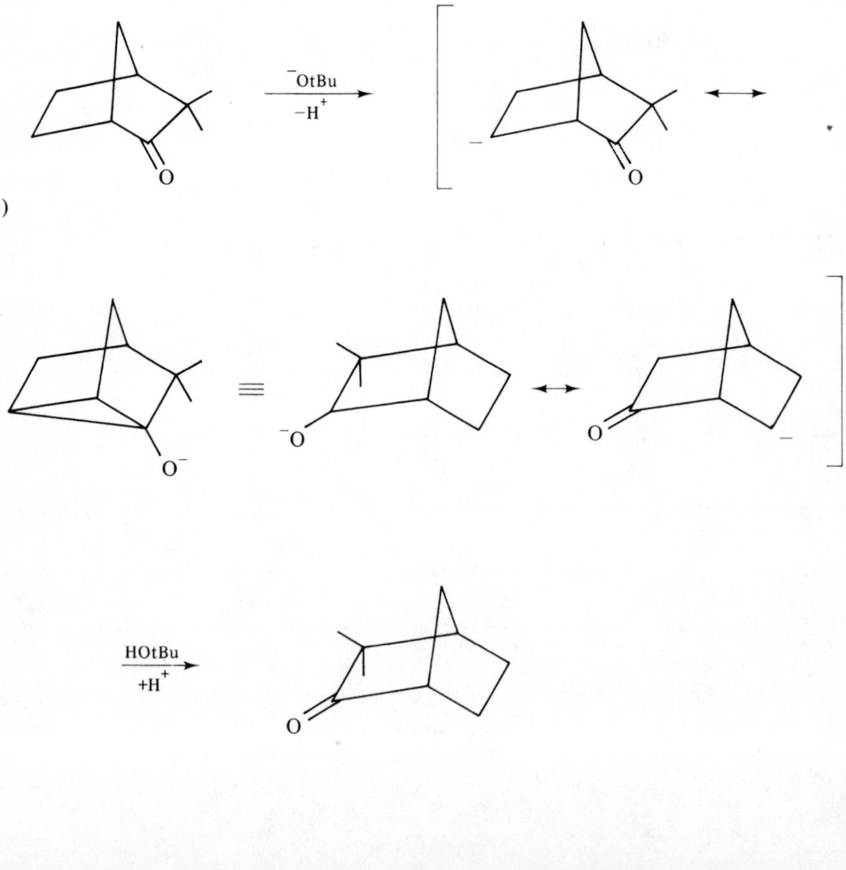

$\xrightarrow[-H^+]{^-OtBu}$

O

O

$\equiv$

O^-

^-O

O

$\xrightarrow[+H^+]{HOtBu}$

O

The rearrangement of methylmalonyl CoA to succinyl CoA (7.29) requiring adenosyl B_{12} (7.30) as a cofactor was also shown to occur by an anionic mechanism. Isotopic labeling experiments indicate the tagged carbon is inserted as indicated. Rates of isotopic scrambling indicate that hydrogen is removed from the methyl group, becomes one of three equivalent hydrogens on the coenzyme, and is returned to the substrate. This is in accord with the generation of a powerful base from the coenzyme and possibly with binding the conjugate base of the substrate to cobalt through carbon. The rearrangement itself occurs through a nonclassical carbanion. Such an intermediate has been demonstrated to account for the racemization of camphenilone by base (7.31). Using an appropriately designed cobaloxime, this ionic mechanism was demonstrated (7.32) for a coenzyme B_{12} analogue.

(7.32)

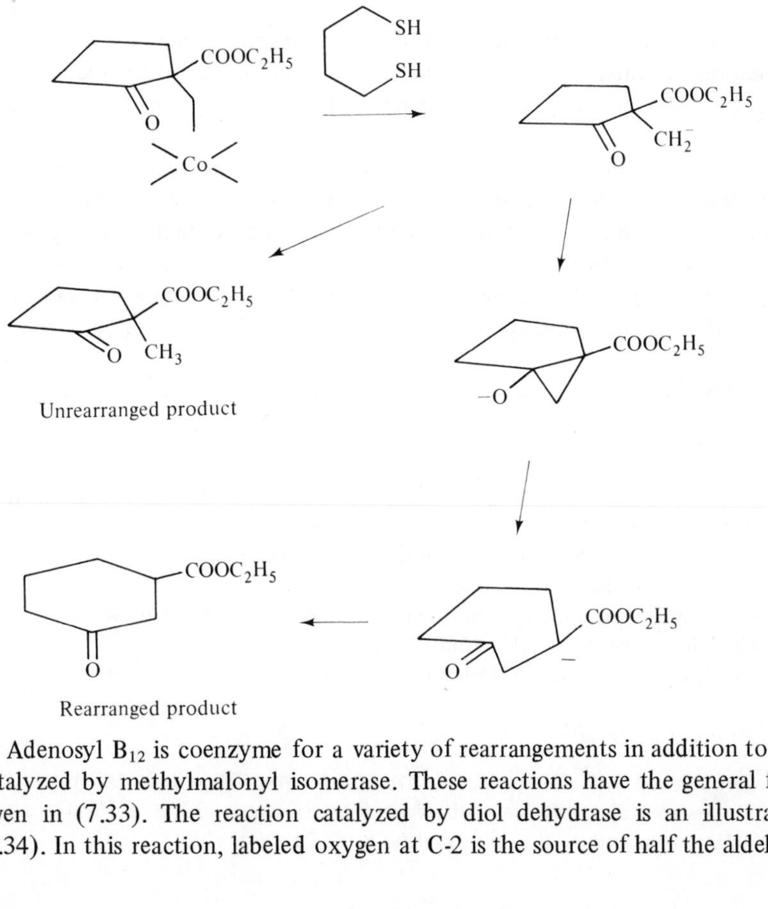

COOC$_2$H$_5$ O   Co   SH   SH   COOC$_2$H$_5$ O   CH$_2^-$

COOC$_2$H$_5$ O   CH$_3$

Unrearranged product

COOC$_2$H$_5$ $^-$O

COOC$_2$H$_5$ O

COOC$_2$H$_5$ O $-$

Rearranged product

Adenosyl B_{12} is coenzyme for a variety of rearrangements in addition to that catalyzed by methylmalonyl isomerase. These reactions have the general form given in (7.33). The reaction catalyzed by diol dehydrase is an illustration (7.34). In this reaction, labeled oxygen at C-2 is the source of half the aldehyde

(7.33)

$$-\overset{X}{\underset{|}{CH}}-\overset{Y}{\underset{|}{CH}}- \longrightarrow -CH_2-\underset{Y}{\overset{X}{C}}-$$

(7.34)

$$CH_3CH(^*OH)CH_2OH \longrightarrow \left[CH_3CH_2CH(^*OH)OH \right] \longrightarrow CH_3CH_2CH{=}O^* \text{ 1/2 of original label}$$

oxygen, showing that the reaction is a rearrangement and not simply the elimination of water. A variety of mechanisms have been postulated for these rearrangements. B_{12} chemical mechanisms form an area of active research interest.

To conclude this chapter, let us return to the comparison of B_{12} to iron porphyrins. Cobalt is required rather than iron, for the d_8 configuration of Co (I), and the long bond to carbon in alkyl cobalamines. Cobalt is not stable in a porphyrin ring, hence the need for the corrin ring. Finally, coenzyme B_{12} provides to biological systems the useful reactions of carbanion chemistry associated with Grignard reagents.

REFERENCES

Ligand Field Theory

F. Basolo and R. C. Johnson, *Coordination Chemistry,* W. A. Benjamin, Inc., New York, 1964.

F. Basolo and R. G. Pearson, *Mechanisms of Inorganic Reactions,* 2nd ed., John Wiley and Sons, Inc., New York (1967).

A. L. Companion and M. A. Komarynsky, *J. Chem. Ed.,* **41**, 257(1964).

F. A. Cotten, *J. Chem. Ed.,* **41**, 466(1964).

L. E. Orgel, *Endeavour,* **22**, 42(1963).

Metal Ions as Catalysts

M. L. Bender and B. W. Turnquest, *J. Am. Chem. Soc.,* **79**, 1889(1957).

B. G. Malmstrom and J. B. Neilands, *Ann. Rev. Biochem.,* **33**, 331(1964).

B. G. Malmstrom and A. Rosenberg, *Adv. Enz.,* **21**, 131(1959).

A. S. Mildvan, "Metals in Enzyme Catalysis", in *The Enzymes,* Vol. 3 (3*rd* ed.), ed. by P. D. Boyer, Academic Press, New York (1970), p. 445
H. Sigel and D. B. McCormick, *Acc. Chem. Res.,* **3**, 201(1970).

Carboxypeptidase

E. T. Kaiser and B. L. Kaiser, *Acc. Chem. Res.,* **5**, 219(1972).
W. N. Lipscomb, *Acc. Chem. Res.,* **3**, 81(1970).

Iron porphyrins

W. S. Caughey, *Ann. Rev. Biochem.,* **36**, 611(1967).
D. M. Collins, R. Countryman, and J. L. Hoard, *J. Am. Chem. Soc.,* **94**, 2066(1972).
R. Countryman, D. M. Collins, and J. L. Hoard, *J. Am. Chem. Soc.,* **91**, 5166(1969).
E. B. Fleischer, *Acc. Chem. Res.,* **3**, 105(1970).
J. L. Hoard, M. J. Hamer, T. A. Hamor, and W. S. Caughey, *J. Am. Chem. Soc.,* **87**, 2312(1965).
E. Margoliash, *Ann. Rev. Biochem.,* **30**, 549(1961).
H. Theorell, *Adv. Enz.,* **7**, 265(1947).
W. W. Wainio and S. J. Cooperstein, *Adv. Enz.,* **17**, 329(1956).

Catalase

A. S. Brill and H. E. Sandberg, *Biochem.,* **7**, 4254(1968).
S. B. Brown, P. Jones, and A. Suggett, "Recent Developments in the Redox Chemistry of Peroxides," in *Progress in Inorganic Chemistry,* Vol. 13, ed. J. O. Edwards, Interscience Publishers, New York (1970), p. 159.
B. Chance and G. R. Schonbaum, *J. Biol. Chem.,* **237**, 2391(1962).
R. C. Jarnagin and J. H. Wang, *J. Am. Chem. Soc.,* **80**, 786(1958).
R. C. Jarnagin and J. H. Wang, *J. Am. Chem. Soc.,* **80**, 6477(1958).
J. H. Wang, *Acc. Chem. Res.,* **3**, 90(1970).
J. H. Wang, *J. Am. Chem. Soc.,* **77**, 822(1955).
J. H. Wang, *J. Am. Chem. Soc.,* **77**, 4715(1955).

Peroxidase

B. Chance, *Adv. Enz.,* **12**, 153(1951).
T. Iizuka, M. Kotani, and T. Yonetani, *Biochem. Biophys. Acta,* **167**, 257(1968).
P. George, *Biochem. J.,* **54**, 267(1953).
H. S. Mason, I. Onopryenko, and D. Buhler, *Biochem. Biophys. Acta,* **24**, 225(1957).
T. H. Moss, A. Ehrenberg, and A. J. Bearden, *Biochem.,* **8**, 4159(1969).
I. Yamazaki and K. Yokata, *Biochem. Biophys. Res. Comm.,* **19**, 249(1965).
T. Yonetani, H. Schleyer, and A. Ehronberg, *J. Biol. Chem.,* **241**, 3240(1966).

Iron-Sulfur Proteins; Nonheme Iron

B. B. Buchanan and D. I. Arnon, *Adv. Enz.*, **33**, 119(1970).

W. R. Dunham, G. Palmer, R. H. Sands, and A. J. Bearden, *Biochem. Biophys. Acta,* **253**, 373(1971).

J. R. Herriott, L. C. Sieker, L. H. Jensen, and W. Lovenberg, *J. Mol. Biol.*, **50**, 391(1970).

T. Herskovitz, B. A. Averill, R. H. Holm, J. A. Ibers, W. D. Phillips, and J. F. Weiher, *Proc. Nat. Acad. Sci.*, **69**, 2437(1972).

S. J. Lippard, *Acc. Chem. Res.*, **6**, 282(1973).

R. Malkin and J. C. Rabinowitz, *Ann. Rev. Biochem.*, **36**, 113(1967).

E. L. Packer, H. Sternlicht and J. C. Rabinowitz, *Proc. Nat. Acad. Sci.*, **69**, 3278(1972).

W. D. Phillips, M. Poe, J. E. Weiher, C. C. McDonald, and W. Lovenberg, *Nature,* **277**, 574(1970).

L. C. Sieker, E. Adman, and L. H. Jensen, *Nature* **235**, 40(1972).

Vitamin B_{12}

B. M. Babior, T. H. Moss, and D. C. Gould, *J. Biol. Chem.*, **247**, 4389(1972).

H. A. Barker, *Ann. Rev. Biochem.*, **41**, 55(1972).

K. Bernhauer, O. Muller, and F. Wagner, *Angew. Chem.*, **75**, 1145(1963).

J. D. Brodie, *Proc. Nat. Acad. Sci.*, **62**, 461(1969).

S. A. Cockle, H. A. O. Hill, R. J. P. Williams, S. P. Davies, and M. A. Foster, *J. Am. Chem. Soc.*, **94**, 275(1972).

R. G. Eager, Jr., B. G. Baltimore, M. M. Herbst, H. A. Barker, and J. H. Richards, *Biochem.*, **11**, 253(1972).

M. Essenberg, P. A. Frey, and R. H. Abeles, *J. Am. Chem. Soc.*, **93**, 1242(1971).

P. A. Frey and R. H. Abeles, *J. Biol. Chem.*, **241**, 2732(1966).

P. A. Frey, M. K. Essenberg, and R. H. Abeles, *J. Biol. Chem.*, **242**, 5369(1967).

P. A. Frey, M. K. Essenberg, R. H. Abeles, and S. S. Kerwar, *J. Am. Chem. Soc.*, **92**, 4488(1970).

H. P. C. Hogenkamp, *Ann. Rev. Biochem.*, **37**, 225(1968).

A. A. Iodice and H. A. Barker, *J. Biol. Chem.*, **238**, 2094(1963).

R. W. Kellermeyer and H. G. Wood, *Biochem.*, **1**, 1124(1962).

H. Kung and L. Tsai, *J. Biol. Chem.*, **246**, 6436(1971).

L. P. Lee and G. N. Schrauzer, *J. Am. Chem. Soc.*, **90**, 5274(1968).

J. N. Lowe and L. L. Ingraham, *J. Am. Chem. Soc.*, **93**, 3801(1971).

W. W. Miller, and J. H. Richards, *J. Am. Chem. Soc.*, **91**, 1498(1969).

G. N. Schrauzer, E. Deutsch, and R. J. Windgassen, *J. Am. Chem. Soc.*, **90**, 2441(1968).

G. N. Schrauzer and J. W. Sibert, *J. Am..Chem. Soc.*, **92**, 1022(1970).

G. N. Schrauzer, and R. J. Windgassen, *J. Am. Chem. Soc.*, **89**, 143(1967).

F. Wagner, *Ann. Rev. Biochem.*, **35**, 405(1966).

INDEX

A

B

C

D

R

S

T

V